STATES OF MIND

STATES OF MIND

NEW DISCOVERIES ABOUT HOW OUR BRAINS MAKE US WHO WE ARE

Adapted from the original talks by

J. Allan Hobson
Steven Hyman
Jerome Kagan
Eric Kandel
Joseph LeDoux
Bruce McEwen
Kay Redfield Jamison
Esther Sternberg

at the Smithsonian Associates–Dana Alliance for Brain Initiatives lecture series, "Understanding the Human Psyche"

Edited by Roberta Conlan

John Wiley & Sons, Inc.
NEW YORK • CHICHESTER • WEINHEIM • BRISBANE
SINGAPORE • TORONTO

This book is printed on acid-free paper. ♾

Published by John Wiley & Sons, Inc.
Published simultaneously in Canada

Library of Congress Cataloging-in-Publication Data:
States of mind: new discoveries about how our brains make
us who we are / edited by Roberta Conlan; with contributions by J.
Allan Hobson . . . [et al.].
p. cm.
Includes bibliographical references and index.
ISBN 0-471-29963-4 (cloth : alk. paper)
ISBN 0-471-39973-6 (paper : alk. paper)
1. Neuropsychology—Popular works. I. Conlan, Roberta.
II. Hobson, J. Allan.
[DNLM: 1. Psychophysiology. 2. Brain—physiology. 3. Emotions—physiology. WL 103U55 1999]
QP360.U526 1999
612.8′2—dc21
DNLM/DLC
for Library of Congress 98-11719

Printed in the United States of America

10 9 8 7 6 5 4 3 2

Contents

Foreword — vii
David J. Mahoney

Introduction — 1
Roberta Conlan

1 Susceptibility and "Second Hits" — 9
Steven Hyman

2 Born to Be Shy? — 29
Jerome Kagan

3 A Magical Orange Grove in a Nightmare: Creativity and Mood Disorders — 53
Kay Redfield Jamison

4 Stress and the Brain — 81
Bruce McEwen

5 Emotions and Disease: A Balance of Molecules — 103
Esther Sternberg

6 The Power of Emotions — 123
Joseph LeDoux

7 Of Learning, Memory, and Genetic Switches — 151
Eric Kandel

8 Order from Chaos 179
J. Allan Hobson

Notes 201
Index 205

Foreword

What can science tell us about ourselves? These days, the amount of information is unprecedented. It seems that every morning we read of new genes discovered, new explanations about how our minds and bodies work. New treatments for various disorders—and new disorders. Making sense of it all is quite a challenge. Are we the sum of our genes or the result of our childhood experiences? Do our moods spring from our thoughts or from the biochemical goings-on in our brains? Can stress and unhappiness really make us physically ill?

In *States of Mind: New Discoveries about How Our Brains Make Us Who We Are,* eight visionary scientists lead us into a new appreciation of both the "mental" and "biological" aspects of our "humanness." This book began as a series of public lectures, cosponsored by the Dana Alliance for Brain Initiatives and the Smithsonian Associates in 1997 in Washington, D.C. The talks were adapted, with the scientists' help, to make up the chapters you are about to read.

These splendid scientists make us understand how the qualities that define us—our memories and emotions, our ways of coping with situations, the brew of attributes that we think of as personality—can be unique and intangible. And yet, behind these images of ourselves is another dimension. Groundbreaking research, in which these investigators are pioneers, is revealing how our inner lives are brought into being by the tireless, intricately coordinated, biological activity of the brain. Both the intangible and the concrete work together, both influence each other, and both determine our health—or lack of it. As science begins to embrace this relationship, the implications are life-changing.

For everyone who delights in discovering what's beneath the surface, *States of Mind* offers a new way of thinking about the mind and the self. For example, one way to identify ourselves is through our experiences and memories. We learn in these pages how research is revealing the steps by which newly acquired information enters the brain and is transferred into permanent storage in the very cells devoted to memory. Sometimes emotions take precedence; we can all think of someone whose every reaction seems to be an emotional one. This book will explain how a frightening event can follow pathways deep in the brain and emerge, even years later, as an inexplicable feeling of dislike for someone or something.

The research discussed here also gives us a new way to look at illness, both mental and physical. As you will see, even "disorders" can be difficult to separate from the unique life of the mind. But what of problems we deem "physical," like bacterial infection, heart problems, or allergies? Doctors and patients alike have always "known" that stressful circumstances can leave someone more vulnerable to disease. But until recently, the physiology of that phenomenon was unknown. Now the picture is becoming clearer as evidence reveals the interconnections between the brain, the stress response, and the immune system.

When thrown out of balance, this collaboration can produce stress-related illnesses.

As understanding deepens regarding the brain's anatomy and circuitry, as we learn more about the brain's activity that goes into mood, thought, and even personality, it's important to remember that we are more than a series of biochemical exchanges. Why does one fact and not another make the jump into long-term memory? Why can we feel devastated by one crisis yet take another one in stride? These questions are still in the realm of the unknown, and may well stay there. Because of the interpretations performed by our individual minds, the world we live in is entirely our own. By offering glimpses into how these processes work, *States of Mind* takes us deeper into that world.

I would like to thank the Smithsonian Associates for their collaboration with the Dana Alliance for Brain Initiatives in the lecture series that gave rise to this book—a series they have done in each of the past three years to ever-growing audiences. As I write this, the Dana Alliance is approaching its sixth anniversary, an organization of 175 of the nation's leading brain researchers, including six Nobel laureates, all committed to communicating the discoveries of brain research to the public. Two years ago, 65 of Europe's top neuroscientists, including two Nobel winners, formed the European Dana Alliance for the Brain to provide the same information in their countries. And in March 1999, there will be a merging of National Brain Awareness Week, involving hundreds of participating organizations from across the United States who have been involved for the last four years, and European Brain Day, the successful first-ever such celebration in Europe in 1998. We hope you will think of this book as our, and your, salute to that merger—the first

World Brain Awareness Week. Even more, we hope this book will mark the beginning of a never-ending journey of discovery and wonder for you into the marvels of the organ that makes you uniquely you: your brain.

David J. Mahoney
Chairman, Dana Alliance for Brain Initiatives
September 1998

Introduction

EVERY DAY, in the course of ordinary conversation, we use a very small word that we learn early in life. The word is *I.* We say things like "I think so" or "I don't remember" or "I have a headache." And then there's Mark Twain, who wrote, "I have a prodigious quantity of mind; it takes me as much as a week sometimes to make it up."

Beneath the wry joke, Twain was expressing an element of bemused wonder—or exasperation—that many of us might share. Human beings have always wondered, in some fashion, about the self, or consciousness, that seems to inhabit the body we identify as ours and that experiences the world "out there." Who (or what), exactly, is the ubiquitous "I" who so readily thinks, remembers, and feels pain? And where in relation to that "I" is the "mind" that can't be made up?

Although we generally manage to carry on without worrying too much about such philosophical conundrums, most of us, at some point in our lives, have been drawn—if not consumed—by the need to know who we are and to understand how we come by our identity and why we feel and behave the way we do. We question how much we owe to (or can blame on) the set of genes we inherited, to what extent we are the product of the circumstances in which we grew up, and how much is within our own control. When we say of a child, "She's got a temper, just like her dad!" are we reflecting on an innate, inherited characteristic, or on behavior learned from a parent?

Far from being academic, these questions and the answers we seek not only bear on the quality of our relationships with family and friends but also have implications for how we function as a society. To what extent is a bad temper, for example, or an inability to find joy in life, a function of will, and to what extent are they the products of the complex interaction between our genes and our environment? The two debates—over what has been called "the mind-body problem" and over "nature versus nurture"—have engaged philosophers and physicians for centuries.

On the mind-body front, the pendulum has swung first one way and then the other several times. Nearly eighteen hundred years ago, for instance, there was essentially no distinction between the mind and the body. The physician Galen of Pergamum ascribed not only physical health but also psychological temperament, or personality type, and emotional well-being to biological influences—varying concentrations of so-called bodily humors such as blood and phlegm. Galen's approach would be called "holistic" today, the idea being that physicians should attempt to treat the whole person. In the seventeenth century, with the advent of a philosophy known as dualism, this integration of mind and body was sundered. The tenets of dualism were crystallized by the French philosopher René Descartes, who argued that the mind was such an ephemeral phenomenon that it must exist utterly distinct from the obviously physical brain and body. In the nineteenth century, painstaking anatomists countered this notion by showing that disorders of the mind and the emotions arise from injury or other flaws in the physical structure of the brain.

In the past several decades, advances in neuroscience have renewed and clarified the integration of the brain and the mind. We now know, for example, that a number of mental problems, such as obsessive-compulsive disorder and schizophrenia, are associated with structural abnormalities in the brain and are also responsive to treatment by drugs, allowing us to try to alleviate them with biological approaches. And today, although we have yet to unravel the mystery of consciousness, we know that it arises—somehow—from the activity of the 100 billion nerve cells that make up the human brain.

As many neuroscientists like to say, "Mind is what the brain does." We know that if the neurons in certain regions of brain tissue are damaged through accident or illness, we can lose large aspects of our "self"—the ability to make new memories, speak coherently, feel love, recognize faces, or comprehend what music is. By the same token, we also know that the brain is remarkably adaptable—or "plastic," as neuroscientists describe it—able to respond to virtually every experience by modifying its connections. Indeed, without this plasticity, we could learn nothing new. With it, adult victims of stroke, or young children who have had large portions of their brains removed to control dangerous epileptic seizures, can recover and thrive, because their energetic neurons reorganize themselves to take over missing functions.

Most neuroscientists now argue that the biological organ inside our skulls is both source and repository of our elusive identity and of all aspects of cognition and emotion. The balance of chemicals in our individual brains may predispose us to react to life's ups and downs with a characteristic tranquillity or agitation. Disturbances of that chemical balance can trigger mood disorders and mental illness. And burgeoning research into the connection between the brain and the body is reinforcing the idea that the influence flows in both directions—that is, our attitudes and emotions, once regarded as purely a function of "mind," can affect the health of the body, and vice versa. The eight scientists who have contributed to this book are united in their belief that any approach to understanding the human mind must take into account both partners in the dance: It is impossible to separate the function of mind from that of the brain.

It is equally impossible to separate the influences of our genes and our environment, although advocates of one side or the other in the debate over nature versus nurture have often tried to do just that. Proponents of eugenics, a term coined by Sir Francis Galton in the late nineteenth century, argued for the primacy of genetic inheritance and were in favor of improving the human race through selective breeding—an argument that was taken to its horrifying conclusion by Adolf Hitler in the mid-twentieth century. Those who argued for the primacy of nurture tended to proclaim that family and societal influences were fundamentally responsible for everything from general intelligence to mental stability.

Today, however, scientists have repeatedly shown that both influences are at work, especially when it comes to mental illness or substance abuse, as Steven Hyman points out in Chapter 1. The particular array of genes we were born with may make us susceptible to manic-depressive illness or alcoholism, for example, but an environmental trigger, or "second hit," must activate the genes in question in order to make us ill or alcoholic. So although we may not be able to modify our genetic inheritance, Hyman reminds us that the brain is phenomenally responsive to experience. If the brain can learn addiction, for example, it can also be taught to unlearn it.

Genes and environment are also at play in the development of our personality, or temperament. In Chapter 2, Jerome Kagan suggests that we come into the world with a brain chemistry that inclines us to be a bold, "I'll try anything once" sort of person, or a timid watcher from the sidelines, or something in between. But nature is also responsive to nurture: Despite being born with a given temperamental tendency, Kagan notes, a fearless, outgoing child can be traumatized into becoming fearful and hesitant, while support and encouragement can help even the most fearful child grow up to be a poised and sociable adult.

The question of a temperamental predisposition, and its underlying physiology, is crucial to any discussion of the mood disorder known as manic-depressive illness (MDI). In Chapter 3, Kay Jamison describes the prevalence of MDI among gifted artists, writers, and musicians and notes that suicide rates among these individuals are well above those for the general population. In describing many studies that suggest that creativity and mood disorders are somehow related, she raises several questions: If the genes that predispose someone to MDI can be identified, should high-risk individuals undergo genetic testing and gene manipulation? By trying to eliminate the genetic roots of MDI in an effort to rid society of this devastating illness, what else do we risk losing?

Not everyone is subject to MDI's cycle of manic highs and depressive lows. More familiar—but just as dangerous—is the phenomenon that we call stress. As Bruce McEwen explains in Chapter 4, the demands of modern life can chronically overload the physiological "fight-or-flight" system that's designed to help protect us from danger. Chronic stress can not only accelerate a host of illnesses but can also cause damage in parts of the brain that are associated with memory—a direct instance of bodily ills affecting cognitive abilities.

The brain-body connection that links our emotions and our health hinges on certain key molecules that are highlighted in Chapter 5 by Esther Sternberg, who describes new findings that pinpoint the substances at work in the nervous and immune systems. When emotional upsets cause blood levels of the stress hormone cortisol to rise, she explains, the immune system can be shut down, making us susceptible to infection. But too little cortisol can send the immune system out of control, turning the body's defenses against itself. Such findings, Sternberg suggests, make a strong case for the argument that classifying illnesses as either medical or psychiatric is an artificial distinction.

Given that emotions can affect our health so profoundly, it stands to reason that we would benefit from having a clearer idea of how emotions work. But what are emotions, anyway? We all "know"—until we're asked to define it. In Chapter 6, Joseph LeDoux explains his own definition and describes his pioneering research into the biology of one fundamental emotion: fear. Fear plays a crucial role in the formation of emotional memory, whose long-lasting covert effects influence our day-to-day reactions and decision-making ability. Indeed, in describing the workings of the body's fear system, LeDoux notes that its repercussions in memory provide a neurological basis for Freud's theories about the unconscious.

We tend to form indelible memories of events associated with strong emotions such as fear, but we also learn through repetition, a fact that Eric Kandel has put to use in his work on the molecular basis of memory of all kinds. As he explains in Chapter 7, by homing in on the genetic "switch" that triggers the formation of long-term memory from information held in short-term memory, we are gaining a greater understanding of how we remember, why we forget, and what we might someday be able to do about debilitating memory loss.

The formation of memory from what we learn and experience during the day appears to be one of the functions served by our nightly excursions into the surreal territory of dreams, as J. Allan Hobson explains in Chapter 8. As those who remember them can attest, however, dreams are often extraordinarily bizarre. Hobson argues that these strange narratives may be merely the by-products of the brain's nighttime activity, created by the cortex in an effort to make sense of the spontaneous electric storm that takes place in the brain during sleep. Although these fanciful stories have always been a rich source of imagery and inspiration, Hobson discounts the idea of universal dream symbolism. Noting that our dreams are produced by our individual brains, which in turn are the products of a unique blend of heredity and environment, he suggests that dreams can offer each of us meaningful insights into our own psyches and concerns.

As all this research reaffirms, the fundamental characteristics of human consciousness and identity are that they are shaped and reshaped by a brain that is continually adapting to the world around us. Whether we're reading or walking, dreaming or talking, the particular impulses and pathways of the brain's billions of neurons are storing experience, learning and unlearning, and creating us anew in the process. Santiago Ramon y Cajal, the Spanish physician, anatomist, and Nobel laureate, captured the essence of the mysterious nature of the brain's workings when he described its collection of neurons as "the butterflies of the soul." Even as they rearrange themselves with breathtaking plasticity as we grow from infant to child to adult, something ineffable remains that makes us recognizable to ourselves and others from one day to the next throughout our lives.

1

Susceptibility and "Second Hits"

Steven Hyman

In the course of their lifetimes, as many as one in five Americans—regardless of age, race, or sex[1]—will be affected by a major mental illness. These disorders, which profoundly impair thinking, emotions, and behavior, are the product of structural or functional abnormalities in the brain—as real a biological malady as cancer or heart disease. In recent decades, neuroscience has made substantial progress in identifying some of the ways in which the brain's biology goes awry: imbalances in brain chemistry or circuit function, for example, or structural anomalies. But understanding how brain abnormalities arise remains a difficult challenge. Why does schizophrenia or manic-depressive illness strike some members of a family and not others? Can those who remain healthy be assured that their children will also be free of the disease? The short answer is, No one can say for certain, one way or the other. The most scientists can offer are statistical probabilities—a 3 percent chance, or a 14 percent chance, or a 50 percent chance that a child will become ill—depending largely on family history.

As Dr. Steven Hyman, director of the National Institute of Mental Health, explains in this chapter, the interplay of genes and environment in the onset of mental illness is extremely complicated. Mental disorders are probably the product of the interaction between several genes that confer vulnerability to a given disease; the more genes are involved, the harder it is to detect any one of them and to unravel its precise role. Equally problematic is the task of identifying possible environmental "second hits"—nongenetic factors that convert a genetic susceptibility into full-blown illness. Is it something that occurs in the womb, the result of maternal malnutrition, or a bout with a virus? Or is the second hit a trauma that occurs at birth or in early childhood, when the brain is extremely malleable? Although scientists can explain many aspects of how the normal brain functions, much remains unknown. As a result, says Hyman, trying to understand what goes wrong in the brain to produce serious mental illness "may be the most difficult and complex activity that human beings have ever undertaken."[2]

As DAUNTING as the challenge is, there is no more compelling reason to attempt to understand the causes of mental illness than that these various afflictions exact an enormous human cost. The derangements of thought, emotion, and behavior that characterize mental disorders such as manic-depressive illness, depression, schizophrenia, and addiction are agonizing not only for the afflicted individuals but also for their family and friends. The torment of coping with a parent's hallucinations and emotional withdrawal, a sibling's psychotic rage, or a child's self-destructive behavior can exhaust families and leave lasting scars even on those who escape the illness itself. As one woman, who as a child watched both her older brother and her older sister succumb to schizophrenia, said, "They no longer inhabit my present life, but their illnesses haunt me like ghosts."[3]

Part of the torment for family members has long been the uncertainty of knowing whether they or their children might be subject to the same disorder or, in the case of parents of an affected child, whether they could have done something to prevent it. Although physicians as long ago as the mid-eighteenth century recognized mental disorders as illnesses,[4] they could offer little in the way of effective treatment. With no understanding of the causes of irrational or violent behavior, society was more likely to react with suspicion and fear than with compassion. To this day, many people living with the devastating hopelessness of clinical depression, for instance, are still ashamed to seek help. Families try to hide a loved one's schizophrenia or downplay delusional symptoms as mere "eccentricities."

Since the mid-1950s, however, progress in the fields of psychiatry, neuroscience, biology, and genetics has begun not only to remove the stigma that was once attached to these illnesses but also to help produce better treatments for those who are ill. Gradually, the public has come to recognize that mental disorders are the result of something gone wrong in a critical organ of the body: the brain. Thanks to modern brain-imaging techniques such as structural magnetic resonance imaging (MRI), positron-emission tomography (PET), and functional magnetic resonance imaging (fMRI), which reveal regions of the brain that are active under different circumstances, scientists have uncovered subtle and not so subtle abnormalities in brain structure and activity in patients suffering from various mental illnesses. In addition, many years of research have shown that these abnormalities have a strong hereditary component. That is, the risk of developing a mental illness increases significantly if a close family member is affected.

But how do genes cause a defect in the brain? And why does a given illness seem to skip around in a family, affecting one sister but not another? Neuroscientists and geneticists have some answers, but by no means all. To appreciate the scope of the challenge these researchers face, we need first to appreciate the intricacy of the organ whose workings they are attempting to understand.

The human brain is probably the most complex structure in the known universe. At birth, an infant's brain contains about 100 billion nerve cells, or neurons—a quantity that rivals the number of stars in our galaxy. But when we marvel at this complexity, we're not just talking about sheer number of cells. Rather, it's what these cells do. Unlike most other cells in the body—a muscle cell or a fat cell or a liver cell, for example—the neurons of the brain and the central nervous system carry on complex conversations with one another. Each of these billions of neurons makes, on average, several thousand contacts with other cells—and in some cases as many as 200,000. Consider the challenge of talking on the phone with 1,000 or 10,000 people at once and keeping all the conversations straight.

Yet whether we're awake or asleep, our brain cells are doing the neuronal equivalent of a mass phonathon, sending and receiving chemical messages triggered by electrical impulses. They do this by means of specialized appendages. Each nerve cell has a single long fiber called an axon for transmitting information and a fine filigree of fibers called dendrites for receiving information. [Figure 1] The length of a given neuron's axon varies. Some are quite short, but others may extend up to three feet, carrying an electrical impulse from, say, the base of the spine to the tip of the big toe. Three feet may not sound like much, until one imagines the nerve cell as a kite three feet across—with an axon tail that's forty miles long. Within the brain alone, given its billions of brain cells, there are probably about 3 million miles of axons.

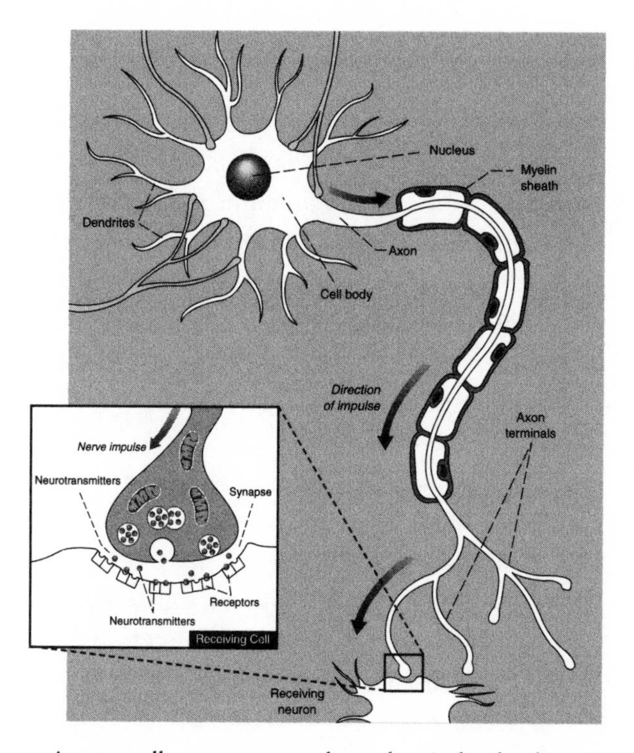

Figure 1 A nerve cell, or neuron, sends an electrical pulse down its myelin-insulated axon to the axon terminals. There chemicals called neurotransmitters are released to float across a small gap, the synapse, to the dendrites of the receiving neuron. If the sum of all incoming signals is sufficient, the receiving neuron will fire, sending an electrical pulse along its own axon to the next neuron in line. Altered from Kibiuk/Society for Neuroscience by Leigh Coriale Design and Illustration. Used with permission.

At its tip the axon splits into terminal regions—sometimes only a few, in other cases as many as several hundred. Each terminal converts the axon's electrical impulse into a chemical one, releasing molecules called neurotransmitters into the tiny gap, or synapse, between it and the receiving neuron. On the receiving end is a mass of fibers called dendrites that emanate from the cell body; each dendrite usually has many branches, each with many receptive zones, allowing each neuron to receive messages from many others. The neurotransmitters—dopamine and serotonin are two of the more familiar ones—float across the synapse to be picked up by specialized receptors, each tuned to a specific neurotransmitter. Any single neuron might communicate using two or three different neurotransmitters, but some are amazingly multilingual; some neurons in the hypothalamus communicate using as many as eight different neurotransmitters. Moreover, researchers have recently discovered that a given neurotransmitter, rather than working in strict "lock and key" fashion with just one or two receptors as previously believed, may work with as many as several dozen or more. So far, for example, fourteen different receptors have been found for serotonin.[5] These myriad brain chemicals may excite or inhibit electrical activity in the next cell down the line, but some have effects far more complex and subtle. Since a given target cell is receiving tens of thousands of these messages at once, it must add them up, in effect. If the sum of the signals exceeds a certain threshold, the target cell will fire, sending electrical impulses along its own axon. At the same time, the incoming signal may trigger changes in the receiving cell itself.

Plasticity and Learning

The brain's wiring and communication system is not only stunningly complex but is also constantly changing in response to the environment. Indeed, scientists have been excited by recent findings on the degree to which neurons in many parts of the brain continue to undergo structural change not just through childhood and adolescence, as was once believed, but throughout life. The good news for those of us who are well past young adulthood is that mental exercise, like physical exercise, may keep the brain supple and fit into our eighth and ninth decades. New experiences, at whatever age, can cause the brain to physically alter its synapses—a characteristic known as plasticity. Indeed, those who compare the human brain to a digital computer do the brain a major disservice. No digital computer comes equipped with an army of lilliputian technicians who climb around and rewire the machine in response to every environmental stimulus.

A key function of some of this rewiring of the brain is learning. Most of us, for example, if prompted with a date like 1776, can probably dredge up "Declaration of Independence!" Some of us, if asked the date of the Norman Conquest, can instantly reply, "1066." These facts were drummed into us somehow by our elementary-school teachers, and we've carried them around for decades. Now, how is that possible? And how is it possible that someone who hasn't been on a bicycle in years can get on one today and still know how to ride? How are "Norman Conquest—1066" and "how to ride a bike" stored?

When we learn facts about the world, or when our bodies learn how to ride a bike or play tennis, our brain is literally remodeling synaptic connections to store the information. This process may involve adding or pruning synapses, strengthening or weakening existing ones. Investigators using techniques such as PET and fMRI have seen wholesale changes in the pattern of brain activity in people who are trained to perform new motor tasks.

If all this remodeling occurs in response to the environment, or "nurture," where does "nature" come in? It turns out that in order to understand how this neuronal transformation occurs—and it occurs all the time as we go about our daily lives—we have to adjust our focus from the level of neurons down to the level of genes. Within each neuron, as with all other cells in the body, is a nucleus that contains an individual's genetic material. Genes determine not only how the brain is built. They also supply the recipes for how its architecture can get rearranged throughout life.

Humans have about 80,000 genes, divided among the 23 pairs of chromosomes. (A full set of chromosomes, all of our inheritable traits, is called the genome.) Chromosomes are long molecules of deoxyribonucleic acid, or DNA, the famous double helix; DNA encodes the information to construct a human being in a simple alphabet made up of molecules known as nucleotide bases, whose names are abbreviated as T, C, G, and A.

Each gene, on average, is several thousand nucleotide bases long and contains the information to make a single protein. Indeed, that is the major function of genes: to instruct the cell in the manufacture of proteins. Proteins, in turn, are critical building blocks of cells. Receptors for the various neurotransmitters are proteins. Some neurotransmitters themselves are small proteins. Enzymes, the molecules that control all the chemical reactions in cells, are proteins. Proteins called growth factors cause nerve cells to grow and sprout dendrites; other proteins work to shrink them. Thus is the very nature of each type of cell in the body determined by its repertoire of proteins.

Given that every cell in your body has the same complement of 80,000 genes, which can make more than 80,000 proteins (because some genes make more than one protein), a fundamental problem during development is to make only the right proteins for the right cells. The cells that produce our hair and fingernails, for example, are expected to produce keratin but not hemoglobin, which is the job of cells in the bone marrow. And we'd prefer that the cells in the midbrain not produce keratin to make fingernails when they should be producing the enzymes to make the neurotransmitter dopamine.

How do the cells know what to do? Through a precise system involving specific sequences of DNA and special controlling proteins, the cells are told which genes can be on and which should be off in any given cell. This is how we get a muscle cell instead of a nerve cell, for example.

During gestation, fetal brain cells multiply rapidly as the brain and the spinal cord assemble themselves. Fascinating research suggests that the fetal brain pulls itself up by its own bootstraps, in effect. Well before there's even a brain as such, these cells begin firing, generating pulsing waves of electrical activity that physically shapes the connections of the brain even as it is still growing.[6] Following orders from perhaps 50,000 genes—more than half the human genome—neural cells begin to lay the brain's foundations, making a kind of "best guess" as to what will ultimately be needed. In the process, they migrate to distant locations to put in place the connections that will link one part of the brain to another. The cerebral cortex, for example, is a structure that comes late in the development of the human brain. The billions of cells that will ultimately mold this outer rind of the cerebrum must somehow push through dense clumps of cells that are already formed[7]—a migratory mass journey akin to having everyone on the West Coast decide to move across the continent.

All this electrical pulsing and neuronal travel are dictated by the genes, which spell out the brain's basic wiring scheme on a kind of macro level, for example, linking the retina to a relay station in the thalamus and the thalamus to the visual cortex. Despite all the information carried in our DNA, however, there's still not enough to produce the final working circuit diagram of our brains. What fine-tunes the precise synaptic connections that nerve cells make with one another is activity produced by the environment.

The Role of the Environment

Even at the earliest stages of development—a one- or two-celled embryo—genes aren't working in a vacuum. They're getting environmental information—a changing supply not only of nutrients but also of instructions—from the cytoplasm of the cell, and from the womb, the uterus. Maternal malnutrition, infections, or drug abuse, for instance, can intrude on the finely orchestrated dance between genes and environment. Indeed, insults to the developing fetal brain are thought to contribute to some forms of epilepsy, mental retardation, autism, and schizophrenia.[8]

Assuming that all goes well in the prenatal environment, a baby comes into the world with its full complement of 100 billion neurons and the appropriate initial connections. But the brain is far from a finished product at this stage. Environmental stimulation continues to be vitally important in the period immediately after birth, and for the first several years of life, in refining and strengthening the still-unfinished blueprint that the genes have laid out. For the human visual system, for example, those environmental experiences consist of visual information. In the case of the development of the emotional circuitry of the brain, we can speculate that early experiences of fear or nurturing will fine-tune certain emotional connections throughout life.

The visual system has been well studied and serves as a dramatic example of the way the environment affects brain wiring.

We know this partly as a result of some tragic "experiments of nature" and partly because of the pioneering animal studies of David Hubel and Torsten Weisel in the 1960s. Occasionally a child will be born with a congenital cataract, a clouding of the lens in one eye that blocks light. If this cataract is removed, the eye will be perfectly normal. But if the cataract is not removed until after the age of three, the child will be blind in that eye, even though the eye itself is optically restored to normal. That is because the wiring of the visual system, the wiring of the connections of the retina to the thalamus and the thalamus to the cerebral cortex, are remodeled by use—by neural firing that causes the release of neurotransmitters.

The first three years of life are a so-called critical period of plasticity for the visual system, a time when these brain circuits have an enhanced ability to respond to environmental information. In order to stabilize synaptic connections in the visual system so that they're retained long-term, the brain needs visual input—light impinging on the retina and activating the release of neurotransmitters, especially a neurotransmitter called glutamate. This activity not only paints a picture of the visual world; it also has an influence on the strength and vitality of the connections in the areas of the brain that are responsible for processing vision. So, in the case of a cataract, if one waits too long beyond the critical period to remove the cataract, the connections from the good eye, which have been stimulated by visual input, literally outcompete the neurons from the blinded eye. They take over nearly all connections to the visual cortex.

We know about critical periods and plasticity in the visual system best of all, but we also know something about it in other primary sensory systems, systems of touch and vibration, and we know a bit about it in the formation of our ability to process and speak language. For instance, we've learned that the age at which you learn a second language determines whether you learn it with or without an accent. Henry Kissinger speaks with a very heavy German accent, but his brother, who is somewhat younger, speaks accentless English. Presumably, when the family immigrated to the United States, Henry had already passed the age of enhanced plasticity for certain motor aspects of learning a language, and his brother had not.

Turning Genes On and Off

How does a stimulus such as light change the way the neurons physically connect to one another? By activating genes, which direct the synthesis of proteins, which, in turn, build or prune synapses. This process of synaptic remodeling occurs not only during brain development but with all learning that produces a long-term memory. Similar processes in different regions of the brain also underlie many responses to psychotropic drugs, brain injury, and illness.

The idea that neural communication itself, drugs, or other stimuli can actually turn genes on or off may seem strange initially, but there is actually an ordinary example from outside the brain that illustrates this phenomenon: the pursuit of physical fitness. Let's say you decide to start working out at the gym; your goal is to build some muscle and get stronger. You begin lifting weights, and after the first workout the major result is that your arms ache for several days. This is because you've stressed the muscle fibers in your arms. But if you lift enough weight, hit the gym enough times a week, and keep at it long enough, eventually you'll have bigger muscles. How does that happen?

The exercise is a kind of stress to the muscle cells. The cell membrane sends a signal to the nucleus of the muscle cell and turns on genes that make muscle proteins; these proteins are needed to respond to the stress caused by the muscles' having to lift unaccustomed weight. Properly carried out, with rest periods for recovery, the repeated stress leads to a stable adaptation to this state of affairs, resulting in increased muscle mass.

As you might expect, such processes are much more complicated in the brain, and produce complex and subtle results, but the principle is the same. Just as muscle cells respond and adapt to signals from the environment, so do brain cells, although in the case of brain cells the signals are mediated by the action of neurotransmitters or, sometimes, by drugs. When neurotransmitters bind with their specific receptor on the outside surface of a cell, a cascade of information flows across the cell membrane into the cell and even to the cell nucleus, with its cache of genes. As described above, each gene has regions containing the information to produce a protein as well as regions that control whether that necessary series of events will occur. (Gene "expression" is the term for the activation of a gene so that the information in its protein coding regions will be "read out," in effect.) So information coming from the environment—information carried by neurotransmitters and then receptors and then all of the intervening signaling steps—can chemically modify some of these control proteins, turning the genes on and off.

Of Mendel and Multiple Genes

So far, we have been describing how our genes are turned on in the right cells at the right time during development and in response to particular environmental stimuli. But we must also address the fact that the versions of each of the genes we inherit from our parents may be subtly different from one individual to another. These differences—a variation in perhaps one or two nucleotide bases in one version of a given gene versus another version of the same gene—contribute to the rich diversity of our species. However, these differences also mean that some of us are more vulnerable than others to illness, including mental illness.

In his original genetic analysis, the Austrian monk Gregor Mendel focused on the traits of pea plants that were each determined by variations in a single gene. Mendel found, for instance, that variation in one gene determined whether a pea plant was short or tall; in another, whether it had yellow peas or green peas; and in yet another, whether the peas were wrinkled or smooth, and so on. Important traits that can be attributed to one gene are thus called mendelian traits.

A number of serious human diseases are mendelian. A single defective gene, for example, can produce cystic fibrosis; another gene variant results in sickle-cell anemia. Among brain diseases, Huntington's disease—which produces abnormal involuntary movements, emotional disturbance, and progressive dementia—is caused by a single abnormal gene, which has recently been identified. The finding of this gene and its protein should begin to provide important clues for appropriate therapy.

When it comes to mental disorders, however, nothing is that simple. We've long known that disorders such as schizophrenia, manic-depressive illness, and major depression tend to run in families, as does alcoholism. Pinpointing whether this susceptibility is the result of shared genes or shared environmental stresses has been difficult, but over the years studies involving identical and fraternal twins have helped us determine the relative contribution made by genes and environment to these disorders.

Identical twins come from the same fertilized egg and share 100 percent of their genes. Fraternal twins, on the other hand, come from two different fertilized eggs and thus, on average, share only 50 percent of their genes, as would any biological siblings. To evaluate the contribution made by heredity, the rate of a given disorder in identical twins is compared with the rate in fraternal twins. If identical twins are significantly more likely to share a disorder, then heredity is probably an important factor. For instance, in manic-depressive illness, if one identical twin is affected, the other has a 60 to 80 percent chance of also having the disorder. A fraternal twin of a manic-depressive individual, by contrast, has only an 8 percent chance of having the disorder. Similarly, the identical twin of someone who has schizophrenia has a 46 percent chance of being affected, whereas a fraternal twin has only a 14 percent chance of being affected.[9]

Adoption studies are also useful in weighing the relative roles of genes and environment. That is, we can ask whether children who were adopted early in life have more in common with their biological or with their adoptive parents. For example, one might have believed that family experiences create the lion's share of vulnerability to alcoholism. However, adoption studies in three Scandinavian countries showed that genes, more than familial environment, influence the risk of someone's becoming an alcoholic. In these studies, adopted sons whose biological fathers were alcoholic were more likely to be alcoholic themselves than were those whose adoptive fathers were alcoholic and whose biological fathers were not. In families where the genetic dice are loaded, so to speak, these genes appear to increase the risk of alcoholism nine- to tenfold over the ordinary sporadic incidence of alcoholism.

Since even identical twins do not always share a disorder, however, scientists cannot emphasize enough that the environment's role in determining whether susceptibility is converted into illness is critical. This complex interaction between multiple genes and multiple environmental factors—the so-called second hits—explains why some families may have several members who are affected with a given disorder, why one child and not another becomes ill, and why the disorder may skip generations.

The idea of second hits is perhaps most familiar in the context of cancer. People who have genes that make them susceptible to cancer may never get it. But if they smoke, that might be an environmental second hit that converts genetic vulnerability into disease. Other people, with different genes, may smoke with impunity—although they may contract a different malady, such as emphysema or heart disease. We also have to keep in mind that a second hit can occur by pure random chance during pregnancy. Building the brain is a very complicated process, and even identical twins manifest many developmental differences. The key thing to remember, though, is that the brain of someone who has a disorder like schizophrenia or depression is functionally and often anatomically different.

The Mechanism of Addiction

We can get a sense of how functional changes in the brain create particular disorders if we look at how drugs of abuse modify the brain to produce addiction. Someone who is ill with alcoholism, for instance, may be past the stage of denial, may fully realize that he is losing a job, friends and family, and his health, may feel terrible and not even enjoy drinking anymore—yet, despite all of these negative consequences, he is unable to control his use of alcohol. In people who are vulnerable to addiction because of genetic or environmental factors, drugs or alcohol alters the way the brain functions, commandeering, in effect, their motivational control.

The key mechanism of addiction resides in a particular chemical system of the brain, involving the neurotransmitter dopamine. There are several dopamine systems in the brain, and they have many jobs. The one that relates to addiction operates between certain areas of the midbrain and the limbic system; it is therefore associated with emotional behavior.

One way of characterizing the job of this dopamine circuit is that it's a reward system. It says, in effect, "That was good, let's do it again, and let's remember exactly how we did it." This reward circuit is so useful that it has been preserved throughout evolution; as a result, we can study it in animal models. Certainly things like sexual reproduction have to be sufficiently rewarding, or nature's experiment with the mechanism would have been a bust.

But even though something like sexual attraction may be hardwired, this circuit also has to be able to learn in order to discover what things in the world are good for the organism. From animal experiments, we've learned that discovery of a highly palatable new food, for example, triggers the release of dopamine in the brain. The pleasant taste was rewarding, and the dopamine circuit makes sure the brain makes a note of it and will remember how to do it again.

Addictive drugs, it turns out, are molecular mimics: They masquerade as neurotransmitters. Coca, or cocaine, in particular, looks enough like dopamine chemically to interfere with the way the brain ordinarily handles the real thing. Normally, dopamine is reabsorbed shortly after its release into the synaptic gap between neurons by the neuron that released it. But cocaine resembles dopamine enough to fool the protein transporters that take up dopamine. The transporters bind cocaine instead, the removal mechanism gets clogged, and dopamine builds up in the synaptic gap. This excess of dopamine makes the cocaine user feel euphoric and extremely alert.

But the price the drug user may pay is a terrible one: addiction. Just as lifting sufficient weight often enough creates adaptations in muscle cells, taking cocaine long enough creates adaptations in the nerve cells. How do they adapt? Dopamine changes the expression of genes in these nerve cells. The activated genes make proteins, which in turn start pruning some connections and strengthening others. As neurons alter their connections in various regions of the brain, the nature of the neural communication between regions is changed.

Moreover, since the drugs have led to a flood of dopamine, the neurons in the brain that normally produce dopamine decrease production for a time. Then, when the cocaine stops coming at the end of a binge, the natural release of dopamine, which is already low, results in sudden deprivation and causes circuits to malfunction. Instead of euphoria, the addict now feels depressionlike symptoms, an inability to take pleasure in the world. This kind of response drives the cycle of renewed drug taking.

Even as drugs directly affect molecules and cells, they also affect certain aspects of learning. Because the dopamine reward circuit teaches the brain, "That was good, let's do it again, and let's remember exactly how we did it," the addict has learned to associate places where he used drugs, and friends with whom he used drugs, with the drug-induced euphoria. Even after addicts have been detoxified, and some of the brain adaptations have been reversed—just as some months after you've stopped lifting weights the muscles shrink—other important changes in the brain don't revert, including this deeply etched learning. This is the root cause of relapses: When people who have been addicted to drugs see the friends with whom they once used drugs, or pass the alley where they used to shoot up, for example, they can suffer intense waves of desire for the drug.

If cocaine takes such advantage of the brain's pleasure and reward system, how is it that not everyone who tries the drug becomes addicted? Cocaine, like all drugs, has side effects. Thus, for some people it can have effects in other circuits in addition to the dopamine reward circuit. Given their particular genetic makeup and predisposition, these people might feel agitated and anxious to such an extent that the unpleasantness outweighs any pleasure they experience. They may therefore be resistant based partly on their own genetic makeup. Or they may, because of environmental factors such as upbringing, recognize the drug's potential for harm and simply decide not to touch it. People who get snared by drugs may not have these kinds of warning signals.

The critical point to remember in all of this is that in the dance of life, genes and environment are absolutely inextricable partners. On the one hand, genes supply the rough blueprint for the brain. Then stimulation from the environment, whether it's light impinging on the retina or a mother's voice on the auditory nerve, turns genes on and off, fine-tuning those brain structures both before and after birth. Genetic predispositions determine, to some extent, which features of the environment an individual responds to, even at an extremely young age. This give-and-take determines our genetic vulnerability to mental disorders or addiction, and whether that vulnerability will lead to illness.

When someone becomes ill, he and his loved ones want first to have the illness identified and then to know whether it can be treated. But the next question is often "How did I get this disease in the first place?" Patients instinctively understand that knowing something about the causes of disease leads to better treatment as well as prevention and perhaps a cure. As researchers learn more about the many forms of mental illness, it becomes easier to test ideas for treatment, which in turn leads to a more sophisticated understanding of the disease itself. Given the breathtaking complexity of the genes/environment dance, mental disorders are considerably more difficult to fathom than sickle-cell anemia and Huntington's disease. The challenges are indeed daunting, but with modern tools and research skills, they are not insurmountable. The search for susceptibility genes in mental disorders will be exciting. Finding them will be life-changing for many.

2

Born to Be Shy?

Jerome Kagan

Monday's Child is fair of face,
Tuesday's Child is full of grace,
Wednesday's Child is full of woe,
Thursday's Child has far to go,
Friday's Child is loving and giving,
Saturday's Child works hard for a living,
But the Child that is born on the Sabbath Day,
Is bonny, and blithe, and good, and gay!

Put to the test, this traditional nursery rhyme would be a poor predictor of a child's personality or temperament. However, its underlying notion—that we are somehow born to be "full of woe" or "bonny, and blithe"—turns out to have some basis in biology. Neuroscientists and psychologists are producing a growing body of evidence that one's predisposition to view the glass as half full or half empty, or to be shy or outgoing, may have biological determinants.

Research conducted during the course of more than two decades by Dr. Jerome Kagan, a professor of psychology at Harvard University and the director of the Mind-Brain-Behavior

Initiative at Harvard University, suggests that our individual brain chemistries bias us, even as infants, to react to the events of life with equanimity or fear. Kagan and a number of other investigators have found striking correlations between, on the one hand, such physiological measurements as heart rate and brain activity and, on the other, observably timid or fearless behavior. These studies offer fascinating clues to the question of how we become who we are. But biology is not necessarily destiny, Kagan emphasizes. Parents, society, and we ourselves have a hand in shaping the way we react to what life throws our way.

WE ALL know them. They may be our children, our parents, our colleagues. We might even recognize ourselves: There the exuberant toddler babbling easily to strangers at the next table in a restaurant, and over here a youngster of the same age and sex crying and clinging to Mom's hand at the door to the kindergarten classroom. There the young adult eagerly accepting a new job in a strange city on the other side of the country, and here the quiet co-worker who becomes tense at any change in routine. Hippocrates called the bold, sociable type sanguine and the fearful, shy one melancholic. Many centuries later, Carl Jung chose the terms *extrovert* and *introvert.* Today, few psychiatrists or psychologists doubt the reality of these two broad personality profiles, although some modern theories of personality call them by different names. However, the precise origins of the many human temperaments—the processes that make for gregarious risktakers or for shy souls who prefer the safety of the known—remain something of a puzzle.

Seventeen hundred years ago, a physician named Galen of Pergamum drew on the then-current theories of human physiology to suggest that personality types were largely the result of a person's inherited physical nature, or constitution, although he also acknowledged that diet and climate could have a moderating influence. Galen believed that humans possessed different concentrations of four fundamental bodily substances, or humors—namely, blood, phlegm, yellow bile, and black bile—and that the combinations of these four humors produced four different temperaments.

The idea of inherited temperaments remained popular from Galen until the end of the nineteenth century but fell out of favor during the first half of the twentieth. At the time, especially in the United States, the political climate tilted toward awarding to environmental influences, rather than genetics, the key determinants of personality.

With the discoveries in the past few decades of the importance of the brain chemicals called neurotransmitters, the pendulum has swung back toward the middle. We might say that we've found the modern analogue of Galen's humors. A number of studies have shown that variations in the concentration of neurotransmitters such as norepinephrine, acetylcholine, dopamine, and serotonin—all of which are vital to the brain's functioning—influence the integrity of the immune system, subjective mood, the ability to concentrate, and vulnerability to mental illness. This research into the myriad effects of neurotransmitters, bolstered by advances in our understanding of how the brain develops, has sparked a renewal of interest in how temperaments arise. If too little or too much of given neurotransmitters can shift our mood from cheerful to morose, investigators theorize, might not variations in neurochemistry affect the developing brain to produce, for example, an inveterate optimist or a confirmed pessimist?

Of course, with more than 100 different molecules at work in our central nervous system, scientists are far from being able to predict exactly which chemical combinations are likely to produce an introvert or an extrovert. Indeed, extreme introverts or extroverts are merely two ends of a broad spectrum. Most of us fall somewhere in between, and in any case there is more to our personalities than simply whether we're shy or outgoing. Future research is likely to reveal a large number of human temperaments, some as rare as that which produces a saintly Mother Teresa, others as familiar and common as the friendly woman in the supermarket checkout line or the curmudgeonly neighbor who complains about the noisy kids next door.

Moreover, any predisposition conferred by our genetic endowment is far from being a life sentence; there is no inevitable adult outcome of a particular infant temperament. As neuroscientists are discovering, the brain is a remarkably adaptable and malleable organ, especially early in life. Even though research suggests that inherited neurochemistries, whatever they may turn out to be, bias young children to react in particular ways—running away from strange people and strange circumstances or embracing the new with enthusiasm—the child's interactions with family, teachers, and peers can shape that predisposition significantly. Whether some event happens willy-nilly, on purpose, or by accident, we learn and change in response to these interactions, to experiences of caring or abuse, even to the experience of, say, a severe childhood illness. By the time a child is only two years of age, his or her temperament is already part of a tapestry whose biological and environmental threads are so tightly woven as to be impossible to tease apart.

Is Your Baby Likely to Be Shy?

Still, as any mother will agree, temperamental tendencies emerge very early, in the first few years of life. Two questions are:

How early do these tendencies emerge? and How stable are they over time? Will most babies who are reactive to novelty grow into shy teenagers and adults? Will the happy-go-lucky four-year-old, for example, still be happy-go-lucky at sixteen? The challenge in studying these phenomena is finding ways of identifying and measuring behaviors that might otherwise be open to subjective interpretations. Indeed, for that very reason, when my colleagues and I began this work we decided not to rely on parental descriptions of their children. First, parents are not equally discerning in their observations; some have difficulty telling the difference between a child who is crying because he's hungry and one who's crying in frustration at being kept in her high chair. Parents also tend to think of their children's personalities in bold strokes. For example, although some infants both cry and smile frequently, most parents tend to focus on only one of these characteristics and describe their infants as either irritable or happy but not both. In evaluating or describing their children, parents also compare one child with a sibling or with other children in the neighborhood. If the firstborn child was extremely irritable, for instance, and the second only slightly less so, the mother will rate the younger child as much less irritable than would objective observers. Finally, different parents interpret the same behavior in different ways. One mother might characterize her child's avoidance of new foods as a sign of her sensitivity, while another might regard the same behavior as reflecting excessive fear.

To bypass the questionable validity of parental descriptions, therefore, we chose to record the behaviors of infants and children directly. From direct observations of more than 500 children over many years, we found two easily observed behavioral profiles, which we call *inhibited* and *uninhibited.* Again, these terms reflect only two of the many possible temperaments. We chose to focus on these two because they are relatively common and are easily distinguished from each other. The inhibited one-year-old, even if he is brought up in an affectionate home without serious trauma, will, by the first birthday, behave timidly and cautiously when he or she is in a strange place or encounters an unfamiliar person or object. This child's initial reaction to novelty is to become quiet, hold a parent's hand, or retreat from the unfamiliar altogether. When asked to meet a new relative, for instance, this child might run and hide in another room. However, once the inhibited child has had a little time to understand or come to terms with the new event, he or she relaxes and the timidity vanishes. The uninhibited child, by contrast, is not excessively shy with strangers or timid in most situations. He or she is not wild or hyperactive but simply spontaneous, laughing and smiling easily.

In setting about our research, we were testing a hypothesis based on the findings of contemporary neuroscience regarding the amygdala, an almond-shaped structure adjacent to the temporal lobe. In the ordinary course of events, information from all the senses—vision, touch, smell, hearing, taste—arrives at the amygdala, which, in turn, communicates with the many parts of the brain that participate in thought, emotion, planning, and behavior. Neuroscientists have discovered that the amygdala is a key player in the human fear response, which is also called the fight-or-flight response. Its job is to evaluate all that sensory information for potential danger and signal other structures, such as the hypothalamus which has as one of its chief duties the monitoring of the production of the stress hormone cortisol. Signals from the amygdala also activate the sympathetic nervous system, which mobilizes the body's responses in times of danger, regulating heart rate, breathing, blood pressure, and skin temperature. Thus when something novel or frightening occurs, such as a sudden loud noise or a narrow escape from harm in an automobile accident, the physiological reactions that accompany the feeling of fright—the racing pulse, breathlessness, goose bumps—are the result, in part, of the activity of the amygdala.

Neuroscientists have also discovered that if the amygdala in an animal is stimulated, the animal will flex and stretch its limbs, arch its back, and tense its muscles. It will also emit distress calls. Human infants, as it turns out, will also display these responses. The central hypothesis of our research, then, was that if a four-month-old infant inherited a neurochemistry that rendered the amygdala very excitable, the infant should flex and extend the limbs and cry when presented with novel stimuli. We named this behavioral profile *high-reactive.* If, on the other hand, the infant inherited a different neurochemistry, one that resulted in a less excitable amygdala, the infant should not flex its arms and legs and should not cry when presented with the same stimulation. We named this profile *low-reactive.* Our prediction was that this tendency to be overly excitable or nonexcitable to novelty would continue, so that by the time they are two years old, more high-reactive infants than low-reactive infants would display the behavioral profile we call inhibited, and more low-reactives than high-reactives would display the uninhibited profile. This, in fact, turned out to be the case.

All the children in our study were healthy and had been born to well-educated mothers who took good care of themselves during their pregnancies. Because one can make an infant fearful through excessive punishment or perhaps through drinking and taking illicit drugs during pregnancy, we eliminated these cases. All of our data came from healthy infants who were the product of healthy pregnancies.

The infants entered the study when they were four months old. They were videotaped throughout a forty-minute battery of tests while resting in a cushioned seat with their mothers nearby. First, the examiner placed heart-rate electrodes on the infant's chest and the mother was asked to look at her infant and smile but not talk. The infant then heard a series of tape-recorded sentences read by female voices. Next, the infant saw a set of colorful mobiles moved back and forth in front of his or her face. A cotton swab dipped in very dilute alcohol was then placed under the baby's nostrils. At this point, the baby heard a female voice speaking syllables at different volumes. A balloon was popped behind the baby's head, and finally, the mother returned to stand in front of her child for a final minute.

In studying the videotapes of the infants' behavior, we noted how often and how vigorously they flexed or extended their limbs and arched their backs, and how often they cried, as well as how often they smiled or babbled. We discovered that 20 percent showed a combination of frequent, vigorous limb flexing and body movement combined with crying when presented with the various stimulus events. The movement was not the random thrashing characteristic of hunger in babies of this age but a rhythmic flexing and extending of limbs that often stopped when the stimulus was removed. On occasions, these infants showed a tense spasticity, an arching of the back, and intense crying, implying that they had been overaroused. This is the behavioral profile of the high-reactive infant. Forty percent of the infants appeared to be calm and laid-back. They occasionally moved an arm or a leg but showed no spasticity or arching of the back, and they rarely cried. This is the profile of the low-reactive infant.

The remainder of the infants fell into two other groups. About 25 percent showed infrequent arm and leg movement but cried, often as a reaction to hearing the human speech. We call this group *distressed.* The smallest group, about 10 percent, showed high movement but did not cry. We call this group *aroused.*

Follow-up Testing

To determine whether the babies would grow up as we predicted—that is, that high-reactive infants would become inhibited children and low-reactive infants would become uninhibited children—we brought the same children back to the laboratory when they were fourteen months old and again at the age of twenty-one months. On each occasion, they underwent a battery of episodes that lasted about ninety minutes. During these sessions, the children encountered a variety of unfamiliar situations, although they were always with their mothers. The critical feature of these situations was that they were unfamiliar to the child.

In making the distinction between inhibited and uninhibited children, we decided that consistently displaying fearful reactions to unfamiliar situations would be evidence of the inhibited profile. Further, we adopted rather strict criteria for deciding whether a child displayed fear during one of our procedures. Many children will become quiet when something unfamiliar happens or appears; a smaller number will show a widening of the eyes and stand in a kind of frozen posture. However, we assumed that the clearest index of intense fear was crying or retreating from the unfamiliar event.

Because we wished to distinguish the small number of children who became highly fearful from the majority, who simply became quiet, we coded each child's reaction as fearful only if he or she cried in response to one of the novel events we had introduced, or if the child failed to approach a person or an object we had designated as unfamiliar. For example, we coded a reaction as fearful if the child cried when her shirt or blouse was lifted to put on heart-rate electrodes or when a blood-pressure cuff was placed on her arm. Similarly, a reaction was coded as fearful if the child cried in response to flashing lights, a toy clown striking a drum, a stranger in an unfamiliar costume entering the room, or the noise of plastic balls rotating in a wheel. In one episode, the examiner showed the child a rotating toy, smiled, and said a nonsense word in a friendly tone followed by the child's name. On the second trial she showed the child a different rotating toy, and this time she frowned when she said the same nonsense word with the child's name and there was a stern quality to her voice. Most children simply looked at the examiner or perhaps furrowed their brows. A small number of them cried, a reaction that was coded as fearful.

Upon studying the videotapes of these follow-up sessions, we found that our prediction had been confirmed. More high-reactives than low reactives showed frequent fear behaviors at both fourteen and twenty-one months. [Figure 2]

When these same children were four and a half years old, they returned to the laboratory so that we could determine whether they had retained any aspects of their original temperaments. Once again, we tested their reactions to unfamiliar events. Of course, four-year-old children do not ordinarily become frightened at the kinds of events that were presented to them at fourteen and twenty-one months. Although it is possible to frighten four-year-olds, most of the events that would do so violate ethical standards. Consequently, we were looking for less extreme reactions that would nonetheless reveal a high- or low-reactive temperament at work. We know, for instance, that each mammalian species has a biologically prepared reaction to novelty or threat. For example, rabbits freeze as a car's headlights come up a driveway; monkeys grimace; cats arch their backs. Humans become quiet. When children or adults feel apprehensive in a novel social situation, they often stop talking, smiling, and laughing. Introverts, for example, have great difficulty initiating

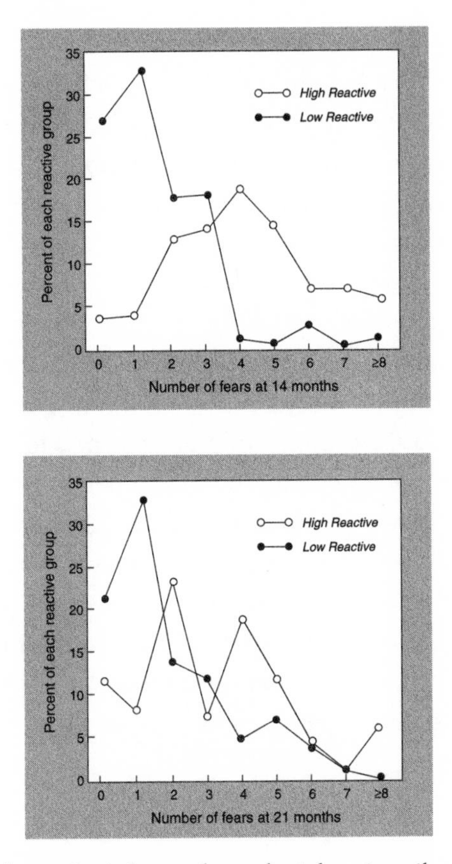

Figure 2 High-reactive infants, whose physiology is easily stimulated, tend to exhibit dramatically more fearful behaviors at later stages of development than their low-reactive counterparts, whose physiology is not as easily aroused. Illustration by Leigh Coriale Design and Illustration from chart by Dr. Jerome Kagan. Used with permission.

conversation at a party with strangers and usually require a long time to relax sufficiently to begin laughing and talking.

In the follow-up at four and a half years, the children were tested by an unfamiliar woman who did not know their prior behavior. The videotapes of each child's behavior were coded for the frequency of spontaneous comments and smiles; a high score indicated that the child felt fairly comfortable and relaxed. The child's answer to an examiner's question was not regarded as a spontaneous comment, but any elaboration on the answer was coded as spontaneous. What we found was that the children who had been high-reactive infants smiled and talked less often than did the low-reactives. Indeed, some high-reactives did not smile or make a single spontaneous comment throughout the session. By contrast, most of the low-reactives talked and smiled frequently.

Several weeks later, these children returned for a second observation in which each child met two other unfamiliar children of the same sex and age, along with the mothers, and played in a large room with toys. Compared with low-reactives, more of the children who had been high-reactive infants were very shy. They were reluctant to interact and stayed close to their mothers for a large proportion of the session. By contrast, those who had been low-reactive infants were sociable and initiated play with the other children.

The Dance of Genes and Environment

It was uncommon for a particular child to show all of the characteristics of the inhibited or the uninhibited profile at every observation. In fact, only 13 percent of the children who had been high-reactive infants were extremely fearful at fourteen and twenty-one months and also very shy and subdued at four and a half years. More than four-fifths of those high-reactive babies, in other words, did not become consistently fearful, but not one became a consistently uninhibited child. This finding means that despite an initial temperamental bias, each child's environment has an important influence on his or her developing profile. As high-reactive children grow older, many decide that they do not want to be fearful and their parents encourage them to overcome their timidity; thus child and family work together to counteract the effects of the initial temperamental bias. Such efforts to cope with fear are clearly effective if the majority of the children who had been high-reactive infants appear to be average four-year-olds. Similarly, low-reactive, laid-back infants, if subjected to trauma or abuse, or even less dramatic environmental stress, can lose their relaxed style. But it was rare for them to become consistently inhibited children.

The temperamental bias doesn't necessarily vanish, however. Underscoring the give-and-take of genes and environment is the fact that it is very difficult to change one's inherited temperamental predisposition completely. Even though children with high-reactive temperaments can learn to overcome their fears, to the point where they appear to be as confident and outgoing as most other children, it is extremely rare for a high-reactive child to show, over the years, the vitality, fearlessness, and emotional spontaneity that is characteristic of most low-reactive children.

The Underlying Biology

Observations of outward behavior provide one source of evidence for ascertaining temperament. Another source is physiological measurements. Given our assumption that the distinctive behaviors of the high- and low-reactive infants were a function of having either a highly excitable amygdala or one that is less excitable, we elected to take measurements of physiological responses that are under the amygdala's influence. So, for example, since the amygdala's signals activate the sympathetic nervous system, which controls heart rate, we expected high-reactives to exhibit higher heart rates in general or larger heart-rate increases in response to some sort of physiological challenge.

When the mothers of the infants were pregnant, they came to our laboratory three weeks before their expected date of delivery. Each woman lay on a recliner as we measured her heart rate. A computer program separated the mother's heart rate, which is low, from the higher heart rate of the fetus. What we found was that the fetus whose heart rate was 140 beats per minute or higher had a high probability of testing later as a high-reactive four-month-old. (Some who had heart rates of less than 140 beats per minute did become high-reactive, but the probability of that occurring was lower.)

We also measured the infants' heart rates when they were two weeks old. We visited the home, waited for the infant to fall asleep, and then measured heart rate at two different times—once when the sleeping infant was supine and again when we asked the mother to pick her sleeping baby up and hold the baby over her shoulder in an erect posture. The erect posture provokes activity in the sympathetic nervous system, raising heart rate and blood pressure, which is why we can get out of bed in the morning without passing out. We found no difference in supine heart rate between the infants who would later become high-reactive and those who would become low-reactive. However, the two-week-old infants who tested as high-reactive three and a half months later had significantly higher heart rates while sleeping in an erect posture than did those who later tested as low-reactive.

It happens that heart rate is also influenced by the parasympathetic nervous system, which is designed to help turn off the fight-or-flight response when the danger has passed. To determine the degree to which heart-rate variability is due to control by the sympathetic nervous system (thus bringing us back to the influence of the amygdala), investigators use an elegant mathematical analysis, called a fast Fourier transformation, which when applied to 60 seconds of heart rate, can, to some degree, estimate the contribution of the sympathetic nervous system to cardiovascular activity. The heart rates of infants who would become high-reactive at four months showed that they were under greater sympathetic nervous-system control than the heart rates of the babies who would become low-reactives.

Another physiological measurement that can distinguish these two temperamental types (high-reactive/inhibited and low-reactive/uninhibited) is the relative neural activity of the left and right hemispheres of the cerebral cortex, particularly in the frontal lobes. This area, just above and behind the eyebrows, is associated with, among other things, such higher mental functions as decision making and working memory, the short-term memory that, for example, allows us to remember a new phone number long enough to dial it. Generally, the level of activity is asymmetrical—that is, greater on one side than on the other. The brain's asymmetry in function was anticipated more than a century ago by Paul Broca's discovery that the left hemisphere participates more fully in speech than does the right. For example, lesions or strokes in the left hemisphere generally impair language function, although people who have suffered damage to their left hemisphere can use their right hemisphere for some aspects of language. In similar fashion, a large body of research indicates that the right hemisphere participates more in states of fear or anxiety than does the left, while the left hemisphere participates more fully in states of joy or relaxation. For instance, older patients who experience a stroke in the right frontal area, resulting in increased dominance of the left hemisphere, report a lighter mood following recovery. If the stroke is in the left frontal area, however, so that the right hemisphere has become more dominant, patients report feeling more tense and worried than they did prior to the stroke. Given this research, we expected that high-reactive, inhibited children would show greater activity in the right than in the left hemisphere, while low-reactive, uninhibited children would show the opposite profile.

One technique for assessing the difference in activity of the two hemispheres is to study the brain-wave pattern called alpha in the electroencephalogram, or EEG. Richard Davidson of the University of Wisconsin has studied the EEG profiles of three-year-olds who had been classified as inhibited or uninhibited and found that the inhibited children had greater activation in the right—less alpha—compared with the left prefrontal area, while the uninhibited children had greater activation in the left frontal area. He has also found that brain-activity patterns tend to remain stable over several years. For instance, children's EEG patterns at age three are good predictors of their level of restraint or exuberance at age seven. When the seven-year-olds were asked to jump to pop bubbles blown over their heads, those with relatively more left-side prefrontal activity hopped vigorously and laughed a lot. Those with more activity on the right, by contrast, were hesitant and restrained, barely rising up on their toes.

Further corroboration of the biological basis for inhibited or uninhibited behavior comes from work done by Nathan Fox of the University of Maryland. Fox tested a large number of four-month-old children using the same battery of novel events that my colleagues and I used. He also recorded the EEGs of these children when they were nine months old and again when they were two years old. Some children were consistently more active in the right frontal area at both later ages; others were consistently more active in the left hemisphere at both ages. A third group of infants were not consistent in their EEG patterns across the two ages. Fox and his colleagues also observed the behavior of the two-year-olds and found that the children who had consistently shown more right-side frontal activity at both ages were the most inhibited; many had also shown high-reactive behavior at four months. The children who were more active in the left prefrontal area at nine months and again at age two were uninhibited at two years of age and were likely to have been low-reactive infants at four months.

Fox suggests that overactivation or underactivation of the two sides of the frontal cortex is associated with differences in what he characterizes as approach and withdrawal behaviors. That is, overactivation of the right frontal cortex should be related to withdrawal and the expression of fearful emotions. Overactivation of the left frontal cortex, by contrast, should be related to exploratory ("approach") behaviors and joyful emotions. Conversely, underactivation of the right frontal cortex should be related to an inability to experience fearful emotions, a condition Fox believes may be observed as an inability to respond to punishment. Underactivation of the left frontal cortex should be related to an inability to experience joyful emotions, a condition that some researchers believe is indicative of depression.

These differences occur in adults as well. Adults who admit to being extremely shy, nervous, or anxious also show greater activation of the right than the left prefrontal area. According to Richard Davidson, adults who exhibit relatively more activity in the left prefrontal area rate themselves as generally more enthusiastic, energetic, and alert. "They say they get more pleasure from life's ordinary activities," Davidson notes.

A final sign of the underlying biology of temperament brings together our hypothesis about the excitable amygdala and the emotional and behavioral differences associated with left- or right-side prefrontal activity. A little-known quirk of the human anatomy is that in the general population both adults and children show a temperature asymmetry on the fingertips of the two hands. Most people's left index finger is cooler by about .15 degrees Centigrade than their right index finger. High-reactive, inhibited children, however, are more likely to have a cooler right index finger compared with the left.

What is the significance of this difference? It turns out that skin temperature in our extremities is controlled by the sympathetic nervous system (we're back to the amygdala again). It also turns out that the sympathetic nervous system, unlike the motor system, is stronger on the same side of the body. That is, whereas movement on the right side of the body is controlled by the left hemisphere of the brain and vice versa, the sympathetic nervous system is more clearly controlled by whichever hemisphere is dominant. If the left hemisphere dominates, that side of the body would have stronger sympathetic influence on the small blood vessels of the fingers; thus when blood vessels constrict to prevent heat loss in cold weather, the index finger on the left hand should be cooler than the index finger on the right. Therefore if high-reactive, inhibited children have cooler right index fingers, the implication is that their right hemisphere—given its association with fear and anxiety states—is dominant.

Gender and Cultural Differences

In Fox's study, as well as our own, there were no sex differences in the relative proportions of boys and girls who were high- or low-reactive at four months. Because the total sample consists of almost 1,000 children across both studies, we can be relatively confident in that estimate. However, by the time children are two years old, more inhibited children are girls and more uninhibited children are boys. I believe this sex difference is a product, in part, of cultural values and family socialization. Most parents are less troubled by fearful behavior in their daughters than in their sons. Moreover, boys in our culture, bombarded by the cultural stereotypes that value "manly" boldness, experience a conflict if they are overly timid and fearful. Thus cultural stereotypes, together with biology, combine to help boys overcome extreme timidity. The story for girls is in some respects just the opposite: The cultural stereotype is for girls to be less boisterous and exuberant, hence the finding that proportionally more girls than boys are inhibited at age two. However, low-reactive infant girls, if they are not otherwise socialized by their families, are more than capable of becoming extremely spontaneous, fearless, uninhibited children. Indeed, one low-reactive infant girl in our sample grew up to be as uninhibited a youngster as any uninhibited boy.

When it comes to cultural differences, there is, unfortunately, not much research on the relations among biology, temperament, and behavior in children from other cultures. Although a few studies suggest that Asian infants are more likely to be low- rather than high-reactive, this does not necessarily mean that Asian infants will become exuberant, uninhibited children. It is not surprising that there might be temperamental differences among groups that have been reproductively isolated for a long time. The relative proportions of the various blood groups, for example, as well as the distribution of some diseases, are different in reproductively isolated populations.

One unexpected biological correlate of high- and low-reactivity shows up in measures of the body type and facial skeleton of the child. Hippocrates, for one, believed that people with a tall, thin, ectomorphic body build were more timid than those with a square, mesomorphic, medium build. (The terms *ectomorph, mesomorph,* and *endomorph* are traditional names for different body types.) Similarly, Freud's monograph on hysteria, written with Josef Breuer, described one woman they labeled hysterical as thin and bony. As part of our study, we measured the facial skeletons of high- and low-reactives and computed the ratio of the width of the face to the length of the face. High-reactive infants, both girls and boys, had narrower faces than did low-reactives—their ratios of width to length were less than .59. They were also more likely than low-reactive children to have an ectomorphic body build and a bit more likely to have very light blue eyes.

These children, as we've seen, have a more active sympathetic nervous system. Why should blue eyes, a narrow face, and an ectomorphic build correlate with a reactive sympathetic nervous system and inhibition? Three weeks after conception, the neural tube—the very beginning of the brain and the spinal cord—folds inward and a group of cells called the neural crest begins to migrate. These cells become, among other things, the nerve cells of the sympathetic nervous system, the melanocytes, which determine the color of the skin and the iris; and the bones of the face. It is possible that the genes that contribute to the different characteristics of these migrating neural crest cells also correlate with high reactivity. Future research will have to determine the validity of that speculation.

Implications

Research on the relationship between the biological underpinnings of temperament and the environmental influence of parents and society holds several implications for the developing behavior of the growing child. Depending on environmental circumstances and events, innate temperamental inclinations can have both adaptive and maladaptive outcomes.

As we've shown, low-reactive, uninhibited children do not become apprehensive easily. Thus if these children live in homes that socialize aggression, and if they play in neighborhoods that do not provide temptations for asocial behavior, they are likely to grow up to choose socially valued vocations that involve taking risks, becoming, for example, surgeons, trial lawyers, airline pilots, corporation presidents, or investment bankers. However, if low-reactive, uninhibited children are raised in environments that permit aggression and also offer opportunities for antisocial behavior, they are at slightly higher risk than their high-reactive, inhibited counterparts would be for developing careers marked by delinquent behavior. I do not believe there are genes for crime, but there are genes for the development of a fearless, risk-taking profile—a profile that would not shy away from crime given a sufficient push from the environment.

The possible outcomes and risks for high-reactive, inhibited children are quite different. On the one hand, for example—and not surprisingly, perhaps—it is rare for these children to show signs of delinquent or even disobedient behavior. Given that these are children whose physiology makes them hypersensitive to events that most other people find innocuous, however, it should also not be surprising that this temperament type is probably at higher risk than most of the general population for anxiety disorders, even though most of them will be able to manage life without having to consult a therapist. The evidence we have collected suggests that about 10 percent of high-reactive, inhibited children will show signs of social phobia when they are teenagers. They will try to avoid parties and any situation that requires them to interact with a large group of strangers. They will tend not to speak up in class and will become very apprehensive when they are required to give speeches.

If we consider the differences in subjective feelings that might be associated with the two temperamental types (and, again, remember that these are only two among many possible temperaments), we can begin to appreciate the challenges that parents and educators face in helping inhibited children to overcome their fears and uninhibited children to channel their energy and exuberance. We can also begin to appreciate the sometimes widely varying reactions of co-workers or adult members of our families to similar events.

Remember that information from our senses arrives at the amygdala before it is communicated to the cortex, the so-called thinking part of the brain, and that the amygdala's job, evolutionarily speaking, is to react to possible danger. So if the amygdala perceives a threat and triggers a bodily response—muscle tension, let's say—sensory feedback from the tense muscles may reach our conscious awareness even before we become conscious of the stimulus that caused the amygdala to react in the first place. Often, we become aware of the tension without being able to identify a particular reason for it.

Now, adults differ in the intensity and quality of this sensory feedback from all parts of the body. We all experience it, to one degree or another, in the course of every day. But it is likely that high-reactive, inhibited children, who have an excitable amygdala and a labile sympathetic nervous system, might experience more intense feedback than their low-reactive peers. Moreover, they are likely to experience it more often. The question is: How do they interpret it? The way people interpret the experience of a high heart rate or bodily tension or unease varies from one culture to another. In Western society, where responsibility is placed on the individual, Americans might interpret this tension as the result of having done something wrong—a mistake or an aloof attitude to a friend. It's easy enough to find some trivial misdemeanor to account for the feeling of tension: we didn't kiss our spouse goodbye in the morning; we didn't read our child a bedtime story; we weren't polite enough to a colleague. Adults—or children—who have inherited a temperament that triggers very intense sensory feedback that arises often, and for no apparent reason, might be overwhelmed by a sense that the world is a dangerous place and that, no matter what they do, they are always somehow at fault. They might write as did Ludwig Wittgenstein, one of this century's important philosophers: "It came into my head today as I was thinking about my philosophical work and saying to myself, 'I destroy, I destroy, I destroy.' I am too soft, too weak, too lazy to achieve anything significant."

It may be that one of the most significant consequences of an individual's temperament is the sensory feedback from the body and how it is interpreted. Frequent interpretations of the feedback as guilt and anxiety, for instance, would increase an inhibited child's inclination to avoid situations that trigger those feelings. However, as we have seen, conscious attempts to change one's behavior can be effective. It works when parents and children collaborate on the effort, and it works when adults and therapists do the same. The world is full of examples of inhibited children who manage their fears and grow up to be valuable, contributing members of society. It is also full of uninhibited children who overcome the risks of growing up in a dysfunctional family living in a violent or dangerous neighborhood. In the end, to ask what proportion of personality is genetic rather than environmental is a bit like asking what proportion of a blizzard is due to cold temperature and what proportion to humidity.

3

A Magical Orange Grove in a Nightmare: Creativity and Mood Disorders

Kay Redfield Jamison

The temperament continuum with shy, inhibited people at one end and outgoing, uninhibited people at the other is a palette varied and broad enough to accommodate most of us. Personal life events such as divorce or death or losing a job might lay us low for a while, causing even the most gregarious and optimistic of us to want to be alone with our sadness or grief. Other events—the birth of a child or falling in love or winning the lottery—might send our spirits soaring for a while, although "soaring" is probably a relative term at the inhibited end of the temperament scale. For the most part, though, as scientists are finding, the passage of several weeks or months finds us back at our usual baseline sense of well-being, be it slightly negative or positive.

But what of the estimated one in five American adults who experience severe mood disorders that plunge them, seemingly irretrievably, into profound depression or cycle them through the dangerous highs and debilitating lows of manic-depressive illness (MDI), wreaking havoc on their families and careers? These disorders do not represent, as they might appear at first glance, merely the extremes of the temperament continuum. Rather, they are another order of phenomenon, albeit one that is also, like temperament, rooted in the brain's biology. In the past several decades, neuroscientists have begun to home in on the genetic markers for a predisposition to depression and MDI, with an eye to eventually finding better treatment, if not eliminating the disorders altogether. In the meantime, researchers are continually refining ways to treat those who suffer from these illnesses, whether with drugs, electroconvulsive therapy (ECT), or magnetic stimulation of the brain.

Although there is no arguing that such treatment is essential to help patients remain productive members of society and in control of their own lives, Dr. Kay Redfield Jamison, a professor of psychiatry at the Johns Hopkins School of Medicine, raises the question of what else may be at stake as we unravel the genetic puzzle of mood disorders. Jamison's research on the prevalence of depression and MDI among the families of gifted artists, writers, and musicians leads her to suspect that the genes that predispose an individual to these illnesses might also confer a predisposition for creativity. And, if this is so, does society run the risk of medicating away or otherwise eliminating one source of great art? Discussing this provocative question, Jamison draws extensively on her research for her highly regarded book, *Touched with Fire: Manic-Depressive Illness and the Artistic Temperament.*

MANIC-DEPRESSION and depression are devastating illnesses. From a clinical point of view, and a human point of view, the most important thing about these disorders is that they are killers. Of the 31,000 suicides that occur in the United States every year—more deaths than result from homicides or AIDS—approximately 60 to 80 percent are associated with depression or manic-depression. Between 15 and 20 percent of people suffering from one of these illnesses commit suicide, compared with an annual suicide rate in the general population of about 1 percent. So any discussion of the relationship of these mood disorders to artistic creativity should not in any way be interpreted as an attempt to romanticize them. The American poet Robert Lowell, who was hospitalized approximately twenty times for acute mania, characterized one of his manic episodes as a "magical orange grove in a nightmare." Both aspects, the orange grove and the nightmare, are important in the lives of the artists we will be looking at here.

Although many people have long suspected that there is a link between genius and insanity, that link was based largely on anecdotal documentation until the advent of a number of systematic studies during the past twenty years. These studies have looked at artists both living and dead in an effort to examine the connection between mania and depression on the one hand and the creative output of renowned artists, writers, and composers on the other. The results have been remarkably consistent. Indeed, the degree of consistency raises larger questions: If there is a relationship between certain kinds of psychopathology and artistic creativity, what are the implications for how artists who are afflicted with these pathologies should be treated? And, given that manic-depression, in particular, is a genetic disorder, what do the implications of the new possibilities in genetic research have for society?

Retrospective Diagnosis

Diagnosing a mood disorder in someone who comes in to see a psychiatrist is one thing. But how does one make a diagnosis, retrospectively, of an artist or a writer who is long dead? The diagnostic process turns out to be in many ways the same. Whether the research takes place during a face-to-face examination or by means of poring through the letters, journals, medical records, and observations of an individual's contemporaries, the first object is to find a clinical description. That is, how did this person describe his symptoms, how did he perceive them, and how did he convey them? The next clue comes from looking at the natural course of the illness, at certain patterns of symptoms, and how they play out over time: How severe are the episodes, how long do they last, and how frequently do they recur? Another piece of the jigsaw puzzle, which in some ways is even more informative in the case of a genetic illness like manic-depression, is family history. In the context of a person's family pedigree, is the individual himself the only one who shows symptoms of mania or depression or psychosis, or does he also have first-degree relatives—parents, brothers and sisters, or children—who have similar patterns of psychiatric illness?

If we look first at the clinical description, we can get a sense of how some of the artists who had these illnesses experienced them by examining their writings. The first of the following three brief descriptions is of melancholy, or depression; the second is of mild mania; and the last one is of severe mania. It happens that people who are later diagnosed with depression often initially go to doctors complaining not so much of a depressed mood as of lethargy—exhaustion, sleep disorders of one kind or another, an overwhelming feeling of tiredness. The Anglo-Welsh poet Edward Thomas, who died in World War I, put it this way: "There will never be any summer anymore and I am weary of everything. I stay because I am too weak to go, I crawl because it is easier than to stop. There is nothing else in my world but my dead heart and brain within me."[1]

Toward the opposite end of the continuum is mild mania. In the beginning, there's a wonderful sense of well-being and of being at one with the universe. The American poet Theodore Roethke, who had manic-depression, wrote: "For no reason I started to feel very good. Suddenly I knew how to enter into the life of everything around me. I knew how it felt to be a tree, a blade of grass, even a rabbit. I didn't sleep much. I just walked around with this wonderful feeling. One day I was passing a diner and all of a sudden I knew what it felt like to be a lion. I went into the diner and said to the counter man, 'Bring me a steak. Don't cook it, just bring it.' So he brought me this raw steak and I started eating it."[2] As it happens, Roethke wrote this just before the dean of his faculty (he was teaching in an English department at the time) had him committed to a psychiatric hospital for mania.

Roethke was describing the initial phase of a particular kind of mania, a visionary, cosmic state. But this state can then develop into a more severe form, a psychotic phase of mania during which people hallucinate and become delusional. Robert Lowell, for example, wrote: "Seven years ago I had an attack of pathological enthusiasm. The night before I was locked up I ran about the streets of Bloomington, Indiana, crying out against devils and homosexuals. I believed I could stop cars and paralyze their forces by merely standing in the middle of the highway with my arms outspread. Bloomington stood for Joyce's hero and Christian regeneration. Indiana stood for the evil unexorcised Aboriginal Indians. I suspected I was a reincarnation of the Holy Ghost and had become homicidally hallucinated. To have known the glory, violence, and banality of such an experience is corrupting."[3]

As Lowell describes, there are several aspects of manic experience. In addition to the psychosis itself, the individual also has a grandiose conviction of his or her own importance, combined with often intense paranoia.

One thing that is frequently underestimated clinically about both severe depression and manic-depression is how long it takes people to recover from a manic or depressive episode. Even though people can get better and function again, the process of recuperating and building oneself back up is very hard, and takes a long time. As Lowell said, "For two years I have been cooling off from three months of pathological enthusiasm. It is as though I had been flayed and had each nerve beaten with a rubber hose."

Manic-Depressive Illness over Time

The average age of onset for manic-depression bipolar illness is about seventeen or eighteen. For unipolar depression—that is, major depression without the manic phase—the average age of onset is up to ten years later, when the individual is in his or her twenties. And for the milder form of manic-depressive illness, called cyclothymia, it's somewhere in between. Manic-depression and depression are episodic in nature; after an episode, at least early in the illness, a person will usually return to baseline functioning. A seasonal pattern is frequently associated with these episodes. If the person is not treated, the episodes tend to worsen over time; they recur more frequently, and often in a more severe form. Untreated, mania lasts from one to three months, and bipolar depression, if left untreated, will last for at least six to nine months on average. These are very long periods of severe psychopathology. Later in life, the intervals between episodes may diminish so markedly that the disease looks like a chronic illness. Finally, suicide is the outcome in up to 20 percent of the people who have severe forms of these illnesses.

Lord Byron

The life of Lord Byron is a good illustration of how manic-depression can unfold in one person's life. (As we'll see later, Byron had manic-depressive illness and suicide on both sides of his family.) Although he first reported having severe depression when he was in his early teens, he had fits of uncontrollable rage even as a child of six or seven—at one point he put a knife to his chest. As he grew older, these rages became much worse.

By the time Byron was an undergraduate at Cambridge University, a certain pattern had already set in. As his primary biographer, Leslie Marchand, notes, by then Byron was already experiencing alternate moods of depression and "reckless indulgence," which is an apt way of characterizing some of his behaviors. During this period, Byron himself wrote: "I've recovered everything but my spirits, which are subject to depression, and I consider myself as destined never to be happy. I'm an isolated being on the earth without a tie to attach me to life."[4]

Now, in some ways this might be a fairly typical example of undergraduate angst. But it's important to note that adolescence is a high-risk period for people who have mood disorders, and what Byron endured was far more than normal adolescent angst. Within a few years, he was writing frequently about his experiences with extremely agitated psychological states and very depressed ones, and about the effect they had on his thinking. Mood disorders not only affect mood and sleep and energy but severe depression can have a tremendously dementing quality. (In fact, in older people it's occasionally difficult to determine whether the diagnosis should be senile dementia or severe depression. When people become severely depressed, they can't concentrate or pay attention; they can't follow the track of their own thoughts, much less of conversations; and they often can't read or watch television.)

What Byron describes is the sense of having his fine-tuning out of balance. "I am growing nervous," he wrote. "I can neither read, write, or amuse myself or anyone else. My days are listless and my nights restless. I don't know that I shan't end with insanity, for I find a want of method in arranging my thoughts that perplexes me strangely."[5]

His depressions became more frequent and much worse. They were often suicidal, and his sister and friends, and later his wife, often feared that he would take his own life. Byron often kept a gun by the side of his bed, and he would walk up and down the hallway with it. "He used to get up almost every night," Lady Byron wrote, "and walk up and down the long gallery in a state of horror and agitation, which led me to apprehend he would realize his repeated threats of suicide."[6]

As it happened, Byron didn't commit suicide by shooting himself. He died at the age of thirty-six, in Greece, while fighting in the Greek Independence cause. He was in exile, having been banished from England for various scandalous activities, but he went to Greece, as he said, to die. He was tired of living, and he increasingly found himself in a state of severe depression. "He declared himself the victim of persecution wherever he went," one of his friends wrote, toward the end, adding that Byron said there was "a confederacy to pursue and molest him, and uttered a thousand extravagances, which proved that he was no longer master of himself." On a different occasion, another friend wrote: "He refused all medicine and stamped and tore all his clothes and bedding like a maniac. One of his friends entered next, but soon returned saying that it would require ten such as he to hold his Lordship for a minute, adding that Lord Byron would not leave an unbroken article in the room. The doctor asked Smith to get Byron to take a pill. Pushing past a barricade Smith found Byron half-undressed, standing in a far corner like a hunted animal at bay. As I looked determined to advance in spite of his imprecations of, 'Baih! out, out of my sight! fiends, can I have no peace, no relief from this hell! Leave me, I say,' and he simply lifted the chair nearest to him and hurled it direct at my head. I escaped as best I could."[7]

Patterns of Productivity

In addition to looking at patterns of illness and illness-related events over time, as we have done briefly with Lord Byron, we can also look at an individual's pattern of productivity. We know that as the illness fluctuates, patterns of energy and thinking also fluctuate. We know, too—from studies of writers, for example—that about half of these individuals are highly disciplined and organized. Whatever quota they've set themselves, whether it's a thousand words a day or two thousand words a day, they produce it consistently most days of the year. But the other 50 percent of writers produce in spurts, often writing much of what they're going to produce for a given year within a few months' time, or even a few weeks' time, interspersed with long periods during which they produce nothing, or very little.

In doing biographical studies, one tries to plot out these periods of high activity to see whether they correspond to other changes that are associated with mania, such as fluctuations in sleep patterns or changes in other kinds of behavior, including spending patterns, sexual patterns, relationship patterns, and so forth. One of the characteristics of the mood elevation that comes with mania or mild mania is that the individuals sleep less and have an abundance of energy; their thoughts move fluidly from one topic to another, and their productivity shoots up. As the manic phase continues, these individuals become irritable, paranoid, and convinced of the correctness of their own ideas. This can contribute to impulsive behavior and poor judgment, which can lead to bouts of recklessly spending large sums of money or becoming involved in questionable sexual liaisons.

One person whose productivity pattern has been examined in this light is the German composer Robert Schumann. Eliot Slater and Adolph Meyer, two psychiatrists in England, plotted Schumann's works across his professional lifetime, and demonstrated that the number of pieces of music he wrote was a function of his dominant mood state during a particular year. So, for example, in those years in which Schumann was suicidal or extremely depressed, according to his letters and journals, he was producing relatively little. And in the years when he was mildly manic or manic he produced a great deal. For the last two and a half years of life, Schumann was in an insane asylum, and he produced close to nothing. [Figure 3]

Support for a retrospective diagnosis of manic-depressive illness can also be found in hospitalization patterns. The course of the manic episodes of Theodore Roethke follows a reasonably typical pattern. The first episode of mania for which he was hospitalized occurred when he was relatively young. Then an unusually long period of several years elapsed before the second episode took place. This is a much longer interval than most people with manic-depression experience, which is not to say that Roethke wasn't manic in between, but apparently the attacks weren't severe enough to require hospitalization. Then the episodes became more frequent, and by the time he died they were almost chronic. A fairly pronounced seasonal pattern is also evident, with the manic episodes occurring most frequently in fall or winter.

Similar hospitalization patterns can be plotted for Vincent van Gogh, Edgar Allan Poe, and many other writers, artists, and musicians. With both van Gogh and Poe, for example, the frequency of psychotic episodes or episodes of severe depression in-

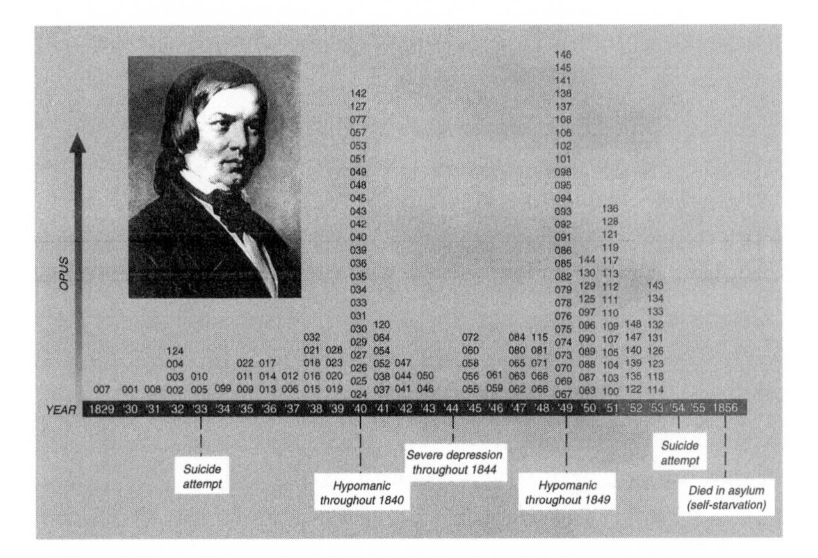

Figure 3 Composer Robert Schumann's productivity, as charted by opus number over nearly three decades, reveals the striking effects of his manic-depressive illness. For example, in each of two years when he experienced manic phases, he created two dozen or more musical works—a stark contrast to the complete absence of musical output during an intervening year when he suffered severe depression. Chart by K. R. Jamison, with permission of S. Karger AG, Basel. Photo: Corbis-Bettmann.

creased over time until their deaths. Van Gogh committed suicide when he was in his mid-thirties. Poe attempted suicide the year before he died of alcohol-related problems, at age forty.

Causes of Death

This brings us to the final aspect of the natural course of this illness, which is: How do people die? There are two major causes of premature death in people who suffer from depression and manic-depression. First and foremost, of course, is suicide. The other major cause is cardiovascular illness. Now, in the general population the suicide rate is about 1 percent—that is, one person in a hundred will commit suicide. According to the few studies that have looked at suicide rates systematically in artists and writers, the rate among these individuals is very much elevated: 6 to 18 percent.

Indeed, the list of artists, writers, and composers who committed suicide—most of them quite young—is staggeringly long. The poet Thomas Chatterton is perhaps the most famous of these young suicides. By the time he died, at the age of seventeen, Chatterton had profoundly affected the work of a number of the Romantic poets. Wordsworth, Byron, Shelley, Keats, and Coleridge all acknowledged the influence Chatterton had on their poetry. Given his youth and his impact, we can only wonder what he would have produced if he had lived longer. Another poet, Vladimir Mayakovsky, one of the great lyric poets of this century, died in his mid-thirties, at almost exactly the same age as Vincent van Gogh, and in the same way—by gunshot. "Just see how quiet the world is," Mayakovsky wrote in his suicide note. "Night has laid a heavy tax of stars upon the sky. In hours like these you get up and you speak to the ages, to history, and to the universe."[8]

The American poet John Berryman, who also committed suicide, is a good example of the tremendous complexity of suicide and its psychological aftermath where remaining family members are concerned. Berryman's father committed suicide, as did his father's sister. Berryman was obsessed with his father's suicide and carried it into much of his work. In one poem, for example, he writes: "I stand above my father's grave with rage,/ often, often before/ I've made this awful pilgrimage to one/ who cannot visit me, who tore his page/ out: I come back for more,/ I spit upon this dreadful banker's grave/ who shot his heart out."[9]

Figure 4 The careworn features of Edgar Allan Poe, who attempted suicide in 1848, reflect the severely agitated depressions that increased in frequency toward the end of his life. Poe died in 1849. Archive Photos/PNI.

Edgar Allan Poe did not overtly commit suicide, although he did take a serious overdose of an opium-based painkiller, laudanum, the year before he died. This photograph [Figure 4] was taken a few days after his suicide attempt. A few days after that, he wrote to a friend: "You saw, you felt the agony of grief with which

I bade you farewell—You remember my expression of gloom—of a dreadful foreboding of ill. . . . I went to bed and wept through a long, long hideous night of despair. . . . The demon tormented me still. Finally I procured two ounces of laudanum."[10]

Lord Byron was often no less tormented, but, unlike Poe, he had a great fund of sardonic wit. "I should, many a good day, have blown my brains out," he wrote, "but for the recollection that it would've given pleasure to my mother-in-law; and, even *then,* if I could have been certain to haunt her. . . ."[11]

Family History

Manic-depression is a genetic illness. That is, it's not just that the individual artist or writer was a particularly sensitive person who felt the assault of the world and therefore grew depressed. Rather, most of the individuals in question had a genetic vulnerability to a mood disorder. More often than not, the illness was evident throughout their families. In addition to this genetic predisposition, then, these individuals had their own inordinately complicated interaction with the world and with their own abilities.

One figure whose family is well documented, on both sides, is Alfred, Lord Tennyson, the Victorian poet. It is useful to track both sides of the family in doing genetic research on manic-depression, because we've found that it's unusual for just one side of the family tree to have the illness. There seems to be a tendency for people with similar temperaments to intermarry, which then leads to an increasing incidence of the illness in their offspring.

Tennyson's pedigree can be tracked back to the late seventeenth century, when the Claytons and the Tennysons intermarried. On the Clayton side of the family there were several indications of mental illness: one person who was declared insane, another who was questionably insane, and another who was imprisoned for violence. On the Tennyson side of the family was one man who hadn't gotten out of bed in two years because of severe depression. We can't say for certain what illness this relative may have had, but it's clear that he wasn't altogether normal.

Eventually, we began to see a significant increase in the number of people in the Tennyson family who were affected, as well as in the severity of the illness. [Figure 5] Lord Tennyson's grandfather had full-blown manic-depressive illness, both of his aunts suffered from depression, and his father was quite insane when he died. The father may or may not have had epilepsy as well, but he almost certainly had psychotic manic-depressive illness. Alfred had eleven siblings, one of whom died in infancy, and most of them had some form of the disorder. One brother died of mania in a hospital after being in an insane asylum for sixty years, and many of the other siblings were either in and out of hospitals throughout their lives or were severely affected by depression and unstable temperaments. Tennyson himself was in treatment for depression at different periods during his adult life. But what is also interesting is that the three eldest Tennyson brothers each walked off with the top literary prizes when they were students at Cambridge.

Lord Byron's family history was filled with all manner of eccentricities and strangeness. On Byron's mother's side of the family—the Gordons, who go way back in Scottish aristocracy—there are at least two suicides. But for thirteen or fourteen generations, there were also instances of murder, violence, and bizarre behavior. "Never was poet born to so much illustrious, and to so much bad blood," one Byron biographer wrote.[12]

On Byron's father's side, it is actually possible to see where the genes for manic-depressive illness may have come into play.

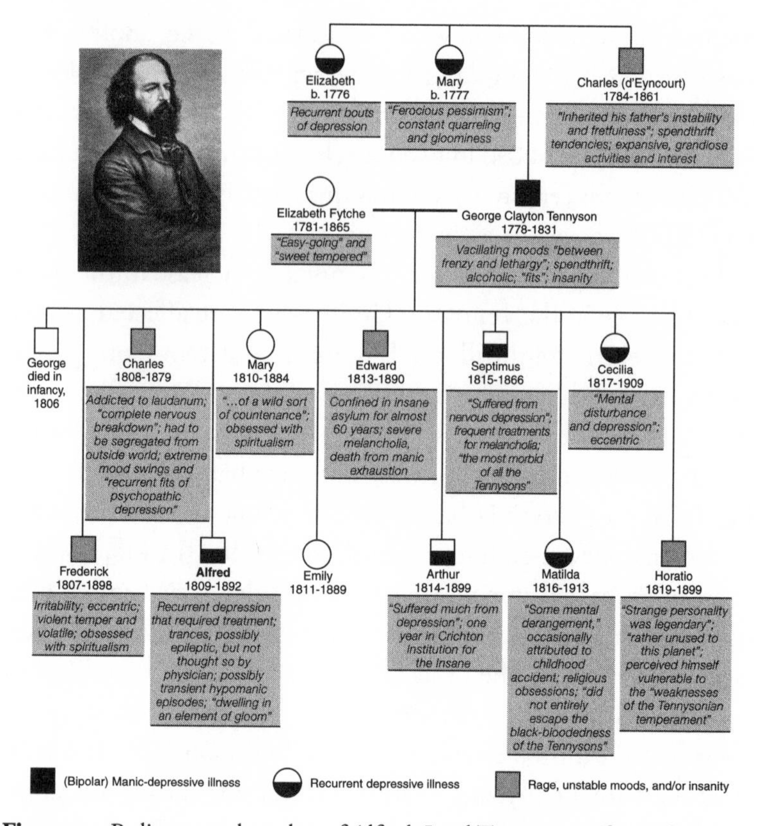

Figure 5 Pedigrees such as that of Alfred, Lord Tennyson, offer evidence that mood disorders such as manic-depressive illness have a genetic basis. Tennyson himself expressed fear of inheriting his family's "taint of blood." Chart by K. R. Jamison, with permission of the Free Press, a Division of Simon & Schuster from *Touched with Fire: Manic-Depressive Illness and the Artistic Temperament* by Kay Redfield Jamison. © 1993, Kay Redfield Jamison. Photo: Archive Photos/PNI.

Byron, the poet, was the sixth Lord Byron. The fourth Lord Byron married into the Berkeley family, which was riddled with manic-depressive insanity. From that point on in the Byron pedigree things become much more deeply disturbed. Byron's grandfather had a debilitating breakdown, and his father, not without cause, was known as "Mad Jack Byron" and was an almost certain suicide. Byron's daughter, Ada, an accomplished mathematician, had grandiose delusions when she was manic, and extremely delusional depressions as well.

Virginia Woolf, who drowned herself, had repeated bouts of psychosis and also came from a family that had a long history of mental illness. Her paternal grandfather had recurrent, severe depressions; both her parents had serious mood disorders; and her first cousin on her father's side, James, died of acute mania. (It was not uncommon, before modern drugs became available, for people to die of mania as a result of sheer exhaustion, systemic infections, or cardiac arrest.) All of Woolf's brothers and sisters were affected with mood disorders, and a niece was also hospitalized for depression.

Ernest Hemingway, as most people know, committed suicide soon after being released from a hospital where he had been treated for psychotic depression. What is less well known is that his father, who was a physician, and had manic-depressive illness, also committed suicide. So, too, did his brother and sister. Two of his sons have been affected with mental illness. One is a physician and has been quite open about his manic-depression. The other son has been treated with shock therapy for repeated psychosis. And, of course, more recently, Hemingway's granddaughter, Margaux Hemingway, committed suicide.

The composer Robert Schumann's family was similarly afflicted. His mother suffered recurrent depressions. His father almost certainly had manic-depressive illness. A sister committed suicide. One son was in an insane asylum for more than twenty years; another son was a morphine addict.

Anne Sexton, the poet, committed suicide. Her sister committed suicide, as did her aunt. There was also a great deal of psychiatric illness, hospitalization, and alcoholism in the family.

Vincent van Gogh has been diagnosed, retrospectively, as having had everything from epilepsy and schizophrenia to digitalis poisoning and porphyria. Less attention has been paid to his family's history of psychiatric problems. Just a few years ago, the Red Cross released records indicating that his brother Cornelius probably committed suicide. A sister, Wilhelmina, was institutionalized for psychosis for thirty or forty years. And his brother Theo, who was an incredible source of support for Vincent, both morally and financially, became psychotic at the end of his life. Indeed, Theo and Vincent wrote back and forth during their lifetime, discussing what they regarded as their shared constitutional malady—namely, melancholia.

Elevated Rates of Mood Disorders

If we summarize some of the studies that have been done in the past two decades, the general finding is a much-elevated rate of depression among writers and artists—eight to ten times the rate of depression in the general population. And the studies that examined mania or mild forms of mania found an exceptionally elevated rate of those disorders as well.

Nancy Andreasen, a schizophrenia researcher at the University of Iowa, was one of the first people to look at the relationship between psychiatric illness and creativity in a systematic way. When she began doing her studies in the 1970s, both the psychiatric community and the literary community thought that in the relationship between so-called madness and creativity, the "madness" was schizophrenia. However, most people who studied schizophrenia found that connection counterintuitive. Schizophrenia is a generally dementing illness, and victims of the disease are rarely successful in getting back to their original baseline functioning.

Andreasen conducted systematic psychiatric diagnostic interviews of the writers in the Iowa Writers' Workshop and found no cases of schizophrenia among them. What she found instead was a remarkably high rate of mood disorders, particularly an extremely elevated rate of bipolar illness, or manic-depression.

I found similar results in the study that I did of British writers and artists. There was an elevated rate of treatment for depression, and—only among the poets—a much elevated rate of manic-depression. These people were getting shock therapy, being hospitalized, or being kept on lithium for manic-depression. In the other groups there were much milder forms of mania, but whatever the form, there was clearly an overrepresentation of mood disorders. Similarly, Hagop Akiskal, at the University of California at San Diego, and his wife, Kareen Akiskal, did a study of European artists and writers and also found an elevated rate of recurrent cyclothymia and depression.

Other researchers have done comparisons of artists and writers with people in other professions. One study, by Arnold Ludwig at the University of Kentucky, compared artists and writers with similarly successful business people, public officials, and scientists. Ludwig's biographical study showed a much higher rate of both mania and suicide in the group of artists and writers than in the group that consisted of public officials, business people, and scientists. The writers and artists group also had six to seven times the rate of involuntary hospitalization for psychiatric reasons. Felix Post, a researcher in Great Britain who looked at lifetime rates of serious depression in male scientists, artists, composers, and writers, found by far the highest incidence of the illness in writers, and the lowest rate, interestingly enough, in visual artists; scientists were somewhere in between.

In addition to comparisons by profession, researchers have looked at rates of mood disorders among male writers and female writers, compared with their nonwriter counterparts. Arnold Ludwig found a greatly elevated rate of depression in women writers over women nonwriters, as well as elevated rates of suicide attempts and mania. Felix Post studied a group of 100 male writers in Britain and found that the combined mood-disorder rate in male writers was extremely high. In the general population, the rate of manic-depression in men and women is just about equal, but the rate of major depression is two to three times higher among women than among men. The rate of depression in the male writers, therefore, is particularly striking.

Given these elevated rates of mood disorders among people who are generally recognized to be highly creative, several investigators have begun trying to sort out how creativity may be related to mental illness. Ruth Richards, a psychiatrist at Harvard, did an interesting study that tackled the question in a new way. First she devised a method of assessing the degree of original thinking required to perform certain "everyday" creative tasks. Then she compared a group of people who had manic-depressive illness, or manic-depressive illness in the family, with a group of normal people and tried to determine which, as a group, was more creative.

She found that the people who had a personal or family history of manic-depressive illness were more creative with regard to day-to-day tasks than those who did not. However—and here's the most interesting finding—within the manic-depressive group, the most creative individuals were the "normal" relatives of the people with manic-depressive illness. This would suggest that there might be some adaptive value in the genes as long as the traits or the illness does not become severe.

But why would there be *any* kind of relationship between these very destructive illnesses and something as extraordinary as artistic and literary creativity? For one thing, the common features of mild mania—an elevation of mood and self-esteem, the need for less sleep, and greater energy—appear to be highly conducive to original thinking. One of the research diagnostic criteria for this phase of the illness is "sharpened and unusually creative thinking and increased productivity." So some people may be creative to begin with, and then, at the onset of a manic phase, the shift in mood and energy level changes their perceptions in a remarkable way. Visionary poets, and the visionary painters, for example, had many of their experiences during phases of acute mania.

Another aspect of the relationship between creativity and mood disorders might stem from the experience of having euphoria alternate with intense psychic pain, from struggling to reconcile the opposite mood states that cause a person to look at the world in one way today and in another way tomorrow. In fact, when artists and writers who have mood disorders are asked what they think is the most important aspect of what happens to them, they will often say that it's the range and the intensity of their emotional experiences—from ecstatic visionary states to despairing, suicidal, melancholic states. There are also periods that could be called subclinical episodes, when people may be neither manic nor depressed but exist in a low-grade variant of these two conditions that may enhance the creative process.

We can't yet say what, exactly, is going on when the brain is depressed, or when the brain is manic, although PET scans illustrate that these are different states, with graphically different metabolic rates. However, we've known for thousands of years that people think differently when they're manic. The premier authority on manic-depression, the nineteenth-century psychiatrist Emil Kraepelin, noted that there was a spontaneous punning and writing of poetry in people who ordinarily had no interest in poetry. It is not uncommon for someone with manic-depressive illness to generate reams and reams of poetry, on paper that's full of colors, underlinings, asterisks, cross-hatchings, exclamation points, and so forth. There's a sense of urgency as well as an enormous productivity. Often, a good deal of rhyming, alliteration, and the use of idiosyncratic words go into manic speech as well.

Some studies have shown that IQ functioning increases during the milder stage of mania, at least up to a point. In addition, in specific word tasks, for instance, people in the manic or slightly manic phase of the disorder can list synonyms or form other word associations much more rapidly than they would normally be able to do. The studies have found that when people are manic or mildly manic, the number of associations goes up threefold, as does the number of associations that are considered to be original. (There are national norms for what people usually say.) Clearly, the brain is activated in a way that is out of the ordinary.

Similarly, a series of very interesting studies in psychology have begun to examine the effect that altering mood has on cognition. In one study, a group of undergraduate students is asked to solve a complicated cognitive task. Half the students have their mood artificially elevated and the other half do not. The result is that the students who had their mood elevated solve the problem not only more quickly but in a more original way.

We've also learned that people who are manic and people who are schizophrenic both suffer severe thought disorder, but the nature of the disorder is very different. In schizophrenia, the thought disorder is highly idiosyncratic, almost autistic. In mania, by contrast, there is a tendency to be much more expansive and to begin combining things. People start combining ideas, relating everything to everything. Sometimes it makes

sense; sometimes it doesn't. Combinatory thought is an important part of creativity, and it is not a characteristic that can be imposed on a noncreative brain to achieve creativity. For a person who is already creative, however, the addition of such a hypercharged state can result in extraordinary combinations.

As we've seen from some of the writings presented here, people who have manic-depressive illness go through life looking at the world very differently from people who do not have the illness. For example, most people who don't have depression or manic-depressive illness don't choose to think about death for any extended period, for obvious reasons. By contrast, individuals who are depressed often have an intimate, moment-by-moment acquaintance with the notion of death; even when they are not overtly suicidal, their thinking is extremely morbid.

Now, the burden that society puts on artists and writers—or, perhaps, the burden that artists and writers are driven to take up—is that of going to the extremes of emotion and experience, and of reconciling many seemingly irreconcilable things in the world. We expect them to look at those aspects of life that we ourselves choose not to look at: to look squarely at how short life is, how decaying the core of the universe is, and how death awaits us all—and then, perhaps, to affirm life in the face of death. So it may be that having manic-depressive illness allows some creative people to achieve those reconciliations.

Implications

This brings us to the issue of implications. If there is a relationship between mood disorders and artistic genius, however one defines it, what risks do we run in treating the disorder or, by means of genetic testing or gene therapy, perhaps in eliminating it altogether? Would we even remember Edgar Allan Poe today if Prozac had been available in the nineteenth century? Many writers and artists—or academics or lawyers or anyone who puts a premium on mental acuity—are concerned about the potential risks of current treatments for manic-depressive illness. They want to know, "What's going to happen to my creativity if I get medicated?" Artists and writers worry that they might lose their facility, their source of inspiration and motivation, the source of their high energy states. Their concern is a reasonable one. The medications that we have for mood disorders work by exerting an effect on those pathways in the brain and the central nervous system that affect mood, energy, sleep, perception—in other words, all aspects of the things that make us human.

So of course there's a risk involved. We know from both animal and human studies, for example, that high doses of lithium can affect the rapidity of thought, energy levels, motivation, and so forth. But we also know that those effects can be mitigated by modifying the dosage. And of course psychiatry is not unique in having the risks of serious side effect associated with treatment. For someone with breast cancer, Hodgkin's disease, or liver problems, the treatments are going to be chemotherapy, radiation, surgery, or other serious medical interventions. Whatever the illness, it is necessary to weigh the risks of treatment against those of receiving no treatment. In that sense, then, psychiatry is no different from any other field.

Moreover, data on the thousands of people who have taken lithium show that two-thirds report having no intellectual or cognitive side effects whatsoever. Two studies actually asked artists and writers, "Are you less productive, more productive, or about the same in your productivity and creativity on lithium as you were before you began taking lithium?" Three-quarters of the artists and writers stated that they felt as productive or more productive. And that's because many of them had been spending a lot of time in psychiatric hospitals when they were manic or severely depressed. So productivity, at least by self-report, is better. Granted, 25 percent of the artists interviewed felt that they were *less* productive. This is not an insignificant finding, and it is an issue that should be addressed by rigorous clinical research. There are now other drugs available, for example, that have fewer side effects and are often used in combination with lithium.

But how do we get people to make meaningful decisions about beginning, and staying on, a course of medication that affects their mood and their very sense of self? Patients who have grandiose and euphoric manias, and who like those manias very much, tend to focus on the risks of treatment; they tend not to focus on the risks of having no treatment. With manic-depressive illness, we have a situation where if the medication is working it's not just taking away pathology, it's taking away certain highly pleasurable, and occasionally highly productive, states as well. As a result, the single most difficult clinical problem in treating manic-depressive illness is keeping people on the medication. The compliance rate is discouragingly low.

We need to focus on the risks of having no treatment. These include, first and foremost, severe depression and the potential for suicide. The fact that the illness, without treatment, is likely to worsen over time makes it more difficult to treat at a later date. It is also more likely that the person will resort to alcohol and drugs, and thereby worsen the course of the illness. So the result of having no treatment is not benign, by any stretch of the imagination. It is rather like allowing strokes or uncontrolled seizures to go untreated and expecting the brain to remain intact. The brain can't continue to experience the imbalances in the chemicals that accompany mania and severe depression without suffering damage.

Finally, we come to the issue of genetics. Genetic research is moving very quickly in manic-depressive illness, and it has very real implications for society. Scientists have already found sites on five chromosomes that may contain the genes that predispose an individual to the illness.[13] So it's not a matter of *whether* the genes will be found but *when.*

The question for society, therefore, is: If we could get rid of these genes or fundamentally alter them, would we, and if so, at what cost to society? On the one hand, the toll this illness takes on the individual is horrific—the pain and suffering, the alcohol and drug abuse, the damage to relationships and jobs, the high suicide rates. And the cost of nontreatment where society is concerned, in terms of violence, alcohol and drug abuse, and hospitalization, is also very high. So, given those costs, the advantages of finding the genes are overwhelming in many respects. With earlier and more accurate diagnosis, we would be able to intervene much earlier in children, before the onset of severe problems. Ultimately, not only could diagnosis become more accurate but researchers could devise better treatments based on the knowledge obtained from the genetic research.

However, as is true of other genetic illnesses, many social, ethical, and legal issues arise in the areas of genetic research and genetic testing. What guarantees are there that the results of genetic tests will remain private? Could test results affect someone's insurance coverage or employment, for example? Moreover, the fact that a person has the genes for manic-depressive illness does not mean he will necessarily become ill. Manic-depression has a variable age of onset, and triggering it involves a complex interaction among light, drugs, sleep, and the environment. Finally, even if illness does strike, it is usually treatable.

Given all these considerations and the fact that research indicates that the genes for this disorder can be advantageous to society—do we really want to tamper with them? Certainly you might not want to have the illness yourself, and you might not want any of your friends or family to have the illness. But as a society we might find it beneficial to know that a certain percentage of the population has the illness—the risktakers who go out and push the envelope. Otherwise, why have the genes for manic-depressive illness survived? Might it be that the decisions that would be best for an individual are not necessarily best for society?

4

Stress and the Brain

Bruce McEwen

The slings and arrows of daily life, whether they are as trivial as a long line at the automatic teller machine or as grave as the news that someone we love has a serious illness, contribute to the modern malady known as stress. Many of us know the signs of our own stress—a pounding heart and rapid breathing, perhaps, or tension revealed in a clenched jaw or throbbing head. All of these are the effects, or aftereffects, of physiological responses that nature designed originally to prepare our bodies to fight or flee from danger. In prehistoric times, these reactions generally subsided once we had either escaped the tiger or killed it. In the modern world, however, the events that trigger our stress response are more often psychosocial anxieties and frustrations than life-and-death battles with a predator. Since neither fighting nor fleeing is an option during a contentious meeting or in the middle of a traffic jam, we may often find ourselves physiologically revved up, as it were, with nowhere to go. As Dr. Bruce McEwen, a professor and the head of the Harold and Margaret Milliken Hatch Laboratory of Neuroendocrinology at Rockefeller University in New York, explains, when our stress responses are chronically kept in overdrive, the effects may not only exacerbate diseases such as hypertension and asthma but may also interfere with the parts of our brains that are involved with memory.

IN 1936, a Hungarian physiologist named Hans Selye pioneered the study of stress—and introduced the concept itself—when he published an article in *Nature* on a phenomenon that he called the general adaptation syndrome. Selye, who carried out much of his work at McGill University and the University of Montreal, was the first person to recognize the paradoxical nature of the body's stress response. On the one hand, the bodily systems triggered by various types of stressors are designed to protect the individual and restore the body's equilibrium, a state known as homeostasis. On the other hand, if these systems are activated for prolonged periods, they also have the potential to damage the body. Indeed, one of Selye's definitions of stress was "wear and tear on the body." Since Selye's groundbreaking work, stress as a word has been adopted—without translation—in virtually every language. People constantly talk about feeling "stressed" or "stressed out," a feeling that seems common around the world.

A good example of the paradoxical effect of the wear and tear of stress occurs in the life cycle of migrating salmon. These creatures must negotiate many obstacles and hazards before they can return to their spawning grounds, and once they have made sure that their genes will be passed on, they die. What they die from, in fact, is too much stress—or, more precisely, too much exposure to their own stress hormones, which come into play to enable them to fight their way upstream. This prolonged exposure causes the fish to stop eating and their immune systems to collapse. The death—from stress, in effect—of one generation of salmon serves a purpose in evolutionary terms, allowing the food supply to be available for the next generation.

The so-called biphasic action of stress hormones—which refers to the way the hormones aid the organism in one phase and have a destructive effect in the other—is evolutionarily useful, or adaptive, for salmon but much less adaptive for modern human society. For one thing, many diseases seem at least to be exacerbated by stress, including coronary heart disease, hypertension, diabetes, gastric ulcers, colitis, and asthma. Although we can't really say for sure that stress per se causes these diseases, it certainly accelerates their course.

One study, for example, examined the death and morbidity rates in Eastern Europe between 1989, following the collapse of communism, and 1993. During this period of great social instability, there was an increase in both mortality and morbidity, the proportion of sickness in the population, in virtually all Eastern European countries. The increased death rate was apparently due in large part to cardiovascular problems and hypertension, and also to suicides and homicides. In Russia, where instability was probably most prolonged, the life expectancy for men has declined from about 64 years to 59 years.

The toll on the individual is obviously high, but stress also takes an economic toll on society. In the United States, for example, if we add up medical care, hospitalization, and lost productivity on the job, the economic cost of stress and stress-related disorders amounts to about $200 billion a year.

However, as Selye himself pointed out, stress is a natural part of human experience. "Stress is not even necessarily bad for you," he wrote. "It is also the spice of life, for any emotion, any activity causes stress." A passionate kiss can be just as stressful as a near-accident in the car—both raise your heart rate, breathing, and blood pressure. Indeed, people tend to talk about good stress and bad stress, and we all recognize that there are many occasions when rising to a challenge is actually good for us. "Good" stress reflects the ability of the body to defend itself and overcome adversity. Adaptation, habituation, and learning how to cope, as well as a more general idea of toughening, are also helpful aspects of good stress: We perform better, and we feel good about it.

In the short run, then, the stress response is a protective survival mechanism, preparing us to take action in the face of perceived danger. However, some aspects of the secretion of stress hormones such as cortisol and adrenaline can lead to damage and to the acceleration of disease. What goes on in the body and the brain during the stress response, and what, if anything, can we do to reduce the negative aspects of stress?

The Fight-or-Flight Response

As described by Selye, the general adaptation syndrome has three phases: an immediate alarm reaction—the so-called fight-or-flight response, which lasts only a few seconds; a continuation of that reaction, called resistance; and the final stage, exhaustion. All three phases are jointly controlled by the two subsystems of the body's autonomic nervous system. The sympathetic nervous system readies the body for action, mobilizing energy resources; the parasympathetic nervous system operates to counteract that response and restore homeostasis.

When the brain perceives a threat, the hypothalamus, which directs the autonomic nervous system, activates the alarm reaction: Signals travel along the sympathetic nervous system to the adrenal glands, which secrete the hormones epinephrine, better known as adrenaline, and norepinephrine into the bloodstream

to be circulated to target organs. Adrenaline gets the heart rate up, for example, and norepinephrine raises blood pressure. These hormones also quicken breathing and inhibit digestion, a nonessential function during a life-threatening emergency or when the body needs to perform at its peak.

If the threat continues beyond a few seconds, the hypothalamus initiates a chain of events—the resistance phase—that prepares the body to sustain the stress response. A pathway called the HPA (hypothalamic-pituitary-adrenal) axis swings into action. The hypothalamus releases a hormone called CRH, or corticotropin releasing hormone, which stimulates the pituitary gland to secrete ACTH (adrenocorticotropic hormone), which in turn stimulates the adrenal glands to produce cortisol. [Figure 6.]

Cortisol has several jobs. It increases the supply of blood glucose to make more energy available, especially to the heart and the brain, and it helps turn fat into energy. It also depresses the reproductive system, which is not an essential function when the body is under siege. Meanwhile, via the sympathetic nervous system, cortisol stimulates immune organs in case the body must deal with injury.[1] However, cortisol also acts in a complex feedback loop to regulate the production of CRH by the hypothalamus and to suppress the immune system in order to prevent it from overreacting to injury and damaging tissues.[2]

Under normal circumstances, when the stress has ended the parasympathetic nervous system shuts off production of the stress hormones. Heart rate, blood pressure, and breathing all return to normal, and the immune system regains its usual vigilance. If the body can maintain this equilibrium, turning the hormones on when we need them and off when we don't need them, we can deal with life's challenges in a healthy way.

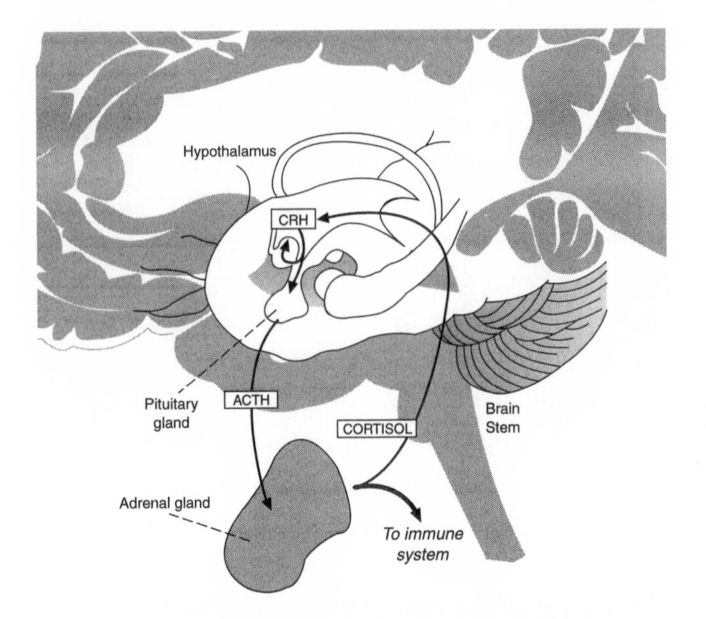

Figure 6 A key phase of the body's response to stress occurs along a pathway called the HPA (hypothalamic-pituitary-adrenal) axis. The hypothalamus releases CRH, a hormone that stimulates the pituitary gland to produce the hormone ACTH, which stimulates the adrenal glands to produce the stress hormone cortisol. In addition to increasing the body's energy supply, cortisol acts in a complex feedback loop to the hypothalamus to regulate production of CRH and to the immune system to prevent an overreaction from damaging tissues. If the HPA axis remains chronically activated, the overload of stress hormones can in turn produce illness. Adapted by Leigh Coriale Design and Illustration, with permission of Roberto Osti.

Allostasis and Allostatic Load

About ten years ago the word *allostasis* was introduced by Peter Sterling and Joseph Eyer of the University of Pennsylvania to describe the process of adaptation to a challenge to the body. *Stasis* means "stability," and *allo* means "variability," so the term refers to the ability to achieve stability through change. Through allostasis, certain systems are turned on by a stressful challenge and achieve stability, or homeostasis, by changing the function of cells. During the time stress hormones are circulating in the body, they bind to receptors on the surface of cells and, either directly or indirectly, activate or inhibit the expression of genes inside the cell nucleus. In this way, stress hormones produce adaptation by determining whether and when various cells manufacture their designated proteins. In effect, stress changes the body's set points, the range within which body systems maintain their operation, much the way a thermostat maintains room temperature. The body's allostatic systems include the autonomic nervous system, the HPA axis, and the cardiovascular, metabolic, and immune systems, all of which protect the organism by responding to stress.

However, when there is a constant external demand, or when these allostatic systems have for some reason been turned on and remain on, their activity stabilizes at a high level, producing an overload, or what is called an allostatic load. One metaphor for this overload might be a seesaw that under normal circumstances is balanced by a five-kilogram weight at both ends. The addition of more weights produces an excess load—wear and tear on the mechanism—even if the weight is distributed and the seesaw remains in balance.

At least three situations can lead to allostatic load. One of them is simply frequent stress. People or animals that are repeatedly challenged in various ways have to keep mounting an allostatic response in order to adapt. Say, for example, that someone is asked to give a talk before a strange group of people. The first time this occurs the person will show an allostatic response: The adrenal glands will produce cortisol, the heart rate will go up, the blood pressure will surge. But usually by the second, third, and fourth time the person gives a talk, the body shows an adaptation or habituation and produces fewer of the physiologic responses that the situation originally aroused. Susceptible individuals, however, whose temperament may make them less self-confident, will fail to habituate; they will continue to produce an allostatic response. As their bodies and brains are repeatedly exposed to stress hormones over the course of weeks or months, they increase the likelihood that there will be wear and tear through allostatic load—thereby increasing the risk of accelerating atherosclerosis (the deposit of plaque on the interior walls of the arteries), for instance, or of having a heart attack.

The second situation is a prolonged response in which the allostatic response is turned on but for some reason fails to be turned off efficiently. For example, people who have a familial history of hypertension tend to prolong their elevated blood pressure even after a challenging situation has passed. Over a long period of time, with many repetitions, this can accelerate the process of atherosclerosis and have other negative effects. The stress of intense athletic training can also induce allostatic load, resulting in reduced body weight, amenorrhea (absence of the menses), and the frequently associated condition of anorexia nervosa.

The third example of a situation that leads to allostatic load, which may seem counterintuitive, is not being able to mount an adequate allostatic response in the first place. When this happens, other systems may get out of control. For example, laboratory-bred animals, called Lewis rats, cannot produce enough cortisol in response to stress. As a result, their systems increase secretion of inflammatory chemicals called cytokines, which are produced by various immune cells and are normally counter-regulated by cortisol. Overproduction of cytokines makes the rats vulnerable to various inflammatory and autoimmune disorders.

To get a sense of how allostasis and allostatic load can produce adaptation or disease, we'll examine three major systems of the body: the cardiovascular system, with its links to obesity and hypertension; the immune system; and the brain and the autonomic nervous system.

The Cardiovascular System

One example of the link among the cardiovascular system, obesity, and allostatic load is Type II diabetes. Individuals with this form of diabetes, whose bodies are resistant to insulin, are caught in a vicious cycle: Their bodies actually produce more insulin, which helps maintain increased production of both cortisol and adrenaline, even as the cortisol acts directly to desensitize insulin receptors, thereby maintaining the body's insulin resistance.

Genetic predisposition is an important factor in Type II diabetes, but because the disease is characterized by increased fat deposition and increased body mass, a person's nutrition, diet, and exercise also play key roles in the onset of the disease. A fat-rich diet increases production of both cortisol and catecholamines, and can by itself lead to insulin resistance and eventually Type II diabetes. Moreover, stressful experiences, by driving up cortisol and catecholamines, can accelerate this process. This is a clear instance of the interaction between nature and nurture, between genes and the environment, with consequences that include hypertension, type II diabetes, obesity, and atherosclerosis.

An illustration of how stress plays into this comes from a study involving monkeys done at Bowman Gray University. When these monkeys are put into groups, they form hierarchies, with a dominant animal and subordinates. When the hierarchy is stable—that is, when the animals have been together for some time and they all know their place—there's no difference between the dominants and the subordinates in the mean thickness of coronary-artery plaque. But if the animals are shuffled into new groupings, the individuals that end up as dominant have to work very hard to establish their place, and they show a marked increase over subordinate animals in the rate of coronary atherosclerosis.

The animal study above echoes the human scenario previously mentioned in Eastern Europe, in which social instability and uncertainty accelerated the disease process. It also points to how differently individuals react to stressful situations, a finding that was corroborated by the Whitehall Studies, which looked at the British civil service during the course of several years. The studies examined the diastolic blood pressure of employees in all six grades of the civil service. What they found was a very regular progression: Blood pressure was lowest among people at the highest grade and highest among those at the lowest grade. It's obviously a very complicated interaction, but something systematic in these people's daily lives is, on the average, affecting their blood pressure. It may be diet, or it may be their life experiences; those at the highest level of the civil service probably have a great deal more control over their circumstances than those at the lowest level, for example. In any case, the study shows that some phenomenon can cause increased blood pressure, which in turn can accelerate the progression toward such illnesses as stroke and atherosclerosis, and increase the risk of heart attack.

The Immune System

The immune system, one of the most fascinating systems in the body, consists of various kinds of cells that are dispersed among a series of immune organs. Immune cells are formed in the bone marrow, then move to the thymus, where the cells are educated to recognize and destroy specific foreign bodies, or antigens. The cells then disperse to the lymph nodes and other immune organs such as the spleen, as well as to such tissues as the lung, the skin, and the liver. Indeed, a key characteristic of immune cells is that they move around: When the body is challenged by something for which there is an immune cell memory, immune cells move to the site.

Dr. Firdaus Dhabhar, a neuroimmunologist at Rockefeller University, has been studying this immune cell movement, or trafficking, and has found that, contrary to popular impression, acute stress enhances rather than impairs this trafficking response of certain immune cells. It's almost as if the body used the stress hormones to improve the movement of these immune cells to their battle stations when they're needed.

Dhabhar studied a particular response called Delayed Type Hypersensitivity, or DTH, which is a kind of overreaction, or hypersensitivity, of tissue that indicates a heightened immune response. In one of Dhabhar's experiments, two laboratory rats are exposed to a simple chemical antigen, so that their systems form an immune cell memory for it; then, when they are challenged with the antigen again, they will have a reaction. At the time of the challenge—which consists of exposing one ear to the antigen—acute stress is produced in one of the animals by putting it in a restrainer so that it cannot move freely. The ear of the rat that was stressed became much redder and more swollen than the ear of the rat that was not stressed, which suggested that the stressed animal's immune system had produced a more extreme inflammatory response to the antigen. Acute stress apparently not only caused the animal to produce more cortisol and more adrenaline for a period of time but also seemed to enhance this Delayed Type Hypersensitive response.

This is surprising, because for a long time the conventional scientific wisdom was that stress normally causes a decrease in the immune cell count in the blood, and that immune cells were probably being destroyed by stress hormones. But in earlier experiments Dhabhar showed that the immune cell count would quickly rebound once the stress had been alleviated, which would be impossible if the cells were actually destroyed. So the question was: Where do these immune cells go? What the experiment involving the chemical antigen in the rat's ear reveals is that the immune cells are not destroyed; instead, they marginate—they go to various tissues and organs and to the inside of blood vessels. When there's a specific challenge, the cells move out of the blood vessels and travel to the site of the challenge. Indeed, an immune cell count in the challenged ear of the stressed rat will be higher than the count in the corresponding ear of the non-stressed rat.

Why would increased inflammation be an advantageous reaction? Dhabhar has suggested that the hypersensitive response under conditions of acute stress might increase the capacity of the immune system to respond to challenges in immune compartments, such as the skin, that serve as major defense barriers for the body. This enhanced vigilance might be especially beneficial in dealing with the effects of vaccination, for example, or in strengthening resistance to infections and such onslaughts as cancer. However, he also points out that a hypersensitive immune response could be harmful in other cases, such as allergies, arthritis, and autoimmune diseases.[3]

Although acute stress can sometimes be beneficial, chronic stress never is. Chronic stress creates a situation of allostatic load. If a pair of laboratory animals are put through the same experiment with the chemical antigen but one animal is subjected to the stress repeatedly for up to five weeks, the Delayed Type Hypersensitivity response is markedly impaired. After one week of stress, the animal is still able to traffic its immune cells out of the bloodstream and into the ear. After three weeks of stress the immune response is smaller, and after five weeks of stress there's very little trafficking response. If the animal is allowed to recover for a week, however, its immune system begins to bounce back, so the damage is not irreversible.

Now, what does this mean for human beings? We know that, in general, human subjects who are in chronic stresslike states—the primary or sole caregivers for Alzheimer's patients, for example—generally show lower DTH responses. This lowered response is also evident in patients suffering from depression, which is often accompanied by elevated levels of cortisol around the clock, and in patients who experience mood disturbances, who also have higher levels of cortisol. There is mounting evidence, too, that people in chronic stress situations, situations of allostatic load, have impaired immune function, whether the situation is a result of external circumstances or is due to factors within their own brains that are keeping these systems active.

The Brain

The brain is the master controller of virtually everything we've discussed so far. It controls the hormones that regulate metabolism. Through the autonomic nervous system, it regulates the activity of the cardiovascular system. It influences the immune system, both through direct innervation, or stimulation by nerves, and through hormones. Most important, perhaps, the brain is the interpreter of what is stressful.

Perception of stress is critical, because what is stressful to one individual may not be stressful to another. The brain has to process an event and decide whether it presents a threat, and, if so, whether the behavioral response will be "fight" or "flight"—or, frequently these days, a "displacement" behavior, such as eating, smoking, drinking (which could make matters worse), or exercise (which may actually be beneficial).

We know that there are individual differences in our reactions to stress that are based on our genetic constitution, as well as on our developmental history and our experiences. All of these factors combine to shape how we apprehend our power to cope with a situation—what we're afraid of, what we're not afraid of, what we feel we can handle.

Now, how is the brain participating in all of this? The brain is not only a controller of many of these actions; it's also, as research with depressed patients indicates, a target for stress hormones. Because of its plasticity—the way its neuronal connections change with experience—the brain itself is vulnerable to the action of stress hormones and can be damaged under extreme conditions.

Two structures in the brain, the amygdala and the hippocampus, buried deep in the temporal lobe, are key players in the interpretation and response to stress. The amygdala is essentially the first structure to be activated in our fear response, and it contributes the emotional flavor to our memories, especially fear-related memories. The hippocampus works closely with the amygdala and is involved in our memory of events and also in spatial orientation. The amygdala is concerned with the emotion of the event; the hippocampus takes care of the so-called contextual, or episodic, memory—the who-what-when-where parts of the story, in effect.

When the hippocampus is impaired, whether this is done by destroying hippocampal cells or by affecting their function, the brain's ability to produce contextual memories is impaired. Research shows that people who have been chronically stressed in various ways do show deficits in memories that are specifically related to the hippocampus and the temporal lobe.

The hippocampus itself has some subsystems and structures, one of which is called the dentate gyrus. The dentate gyrus is involved in a three-way communication with other parts of the hippocampus, a tri-synaptic circuit that is thought to be very important in learning and memory processes. This circuit is the one my colleagues and I have been studying in our effort to learn how the hippocampus responds to stress. One thing we've discovered is that the dentate gyrus is unusual in the adult brain because it continues to produce new nerve cells, a process called neurogenesis. Most nerve-cell formation has stopped by the time we're born, but the dentate gyrus continues to have neurogenesis throughout our adult life. We found this first in mice and rats, and, more recently, in monkeys, so we think it probably happens in the human brain as well. (Assuming it does occur in humans, no one knows what nerve growth late in life actually means, in practical terms; perhaps it simply indicates that the brain continues to hold its own, growing at least some new neurons to replace those that die.)

Even as we were discovering neurogenesis in the dentate gyrus of adult rats, however, we were also finding that stress inhibits this new growth. In rats that have been treated with their own stress hormone repeatedly for three weeks, certain neurons in the tri-synaptic circuit lose many of their dendrites, the fringe of fibers that receive signals from other neurons. Not surprisingly, animals that have been treated in this way suffer from memory loss or impairment.

Similar experiments have been done with the tree shrew, a primitive primate that also makes new neurons during its adult life. If a tree shrew is exposed to a natural stressor—by being put in the cage of a dominant animal, for example—it almost completely stops nerve-cell formation, even if the period of stress lasted only an hour. If the tree shrew is subjected to this single hour of stress every day for twenty-eight days, it loses weight. In other words, the animal is not getting used to the challenge. Each day it is frightened anew, and it produces a lot of cortisol and adrenaline or noradrenaline. At the end of the twenty-eight days, the tree shrew has stopped producing new nerve cells, and its neurons have also lost dendrites, as was the case with rats. So, in both rats and primitive primates, repeated psychosocial stress can result in the atrophy of nerve cells as well as the suppression of the neurogenesis in the dentate gyrus.

Now, one of the things we discovered is that although the stress hormone cortisol plays a key role in this damage to hippocampal neurons, it is not the only culprit. It turns out that a neurotransmitter called glutamate, which is produced within the hippocampus itself, is also involved. Glutamate is an excitatory amino acid, one that triggers a reaction in the next cell down the line, and it is responsible for virtually all of the synaptic transmission that takes place in the hippocampus. So it is an essential chemical in the brain. But if the actions of glutamate are somehow interfered with—with an antiepileptic drug called Dilantin, for example—the stress-induced and even cortisol-induced atrophy of hippocampal neurons is blocked. This tells us that cortisol is working in collaboration with one of the brain's own neurotransmitters to produce these changes. The same thing may be true for the suppression of nerve-cell formation in the dentate gyrus. However, now that we know that a drug like Dilantin interferes with this process, researchers can begin to think about therapeutic interventions.

As these studies and experiments show, the effects of repeated stress on the hippocampus are quite dramatic, although the research also shows that these effects are reversible. In both rats and tree shrews, if the stress stops after twenty-eight days, the dendrites grow back to their normal size. If the stress is considerably prolonged, however, the damage can be permanent. Robert Sapolsky of Stanford University looked at vervet monkeys and found that when subordinate animals have been in a stressful situation for many, many months, they not only lose dendrites in hippocampal nerve cells but they lose the neurons themselves—the neurons are simply destroyed.

Now what about the human brain? Brain scans using functional magnetic resonance imaging (fMRI) of elderly people who show signs of some cognitive impairment (but who don't yet have Alzheimer's disease) reveal that their hippocampi are about 15 percent smaller than normal, as can be seen in the two brain scans below. [Figure 7] The scans also show that the ventricles, the fluid-filled cavities in the middle of the brain, are enlarged compared with those of normal, age-matched subjects.

Indeed, imaging studies show that the hippocampus is one of the first parts of the human brain to show structural changes and a decrease in volume with various stressful conditions. This occurs, for instance, in Cushing's syndrome, a condition in

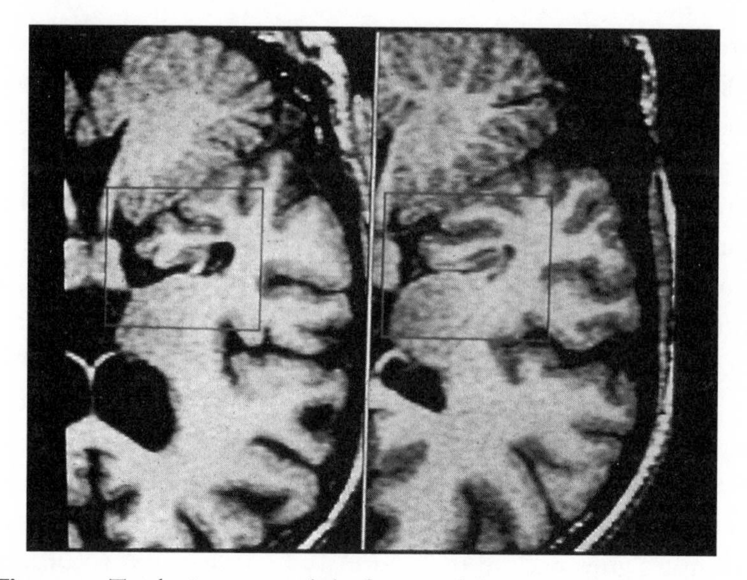

Figure 7 Two brain scans made by functional magnetic resonance imaging reveal that the hippocampus of an elderly person who shows signs of cognitive impairment (right) is about 15 percent smaller than that of a normal age-matched subject (left). Scans courtesy of Dr. Bruce McEwen, Rockefeller University.

which cortisol levels are elevated, probably because of tumors. It also occurs with recurrent depressive illness, and with post-traumatic stress disorder. In addition, hippocampal atrophy occurs in normal aging, preceding dementia, as well as in dementia, or Alzheimer's disease. People who have been diagnosed with schizophrenia also show hippocampal atrophy, perhaps because of the toxic actions of excitatory amino acids as well as stress hormones.

Because there are both reversible and irreversible processes involved, we don't know yet whether these various conditions are due to permanent cell loss or to a potentially reversible atrophy. Some of the data for Cushing's syndrome suggest that hippocampal atrophy may be reversible. If there is reversible atrophy, we might actually be able to treat it with a drug like Dilantin. For recurrent depressive illness or post-traumatic stress disorder, if there is, in fact, permanent cell loss later on, and we could intervene early enough in the development of the disorder, we might slow that process down.

Coping with Stress

Given that stress is very much an individual matter—what's stressful for one person isn't necessarily stressful for another person—what are the factors that determine vulnerability, distress, and allostatic load? Genetic predisposition and early trauma can contribute to a lifelong pattern of emotional responsiveness—that is, a tendency to be hypersensitive to mildly stressful events, or even to events that wouldn't even register on most other people's stress meter. Conversely, some people seem to thrive on stressful conditions. We've all read stories about air-traffic controllers and how stressful their jobs are. However, we've also recognized from reading those articles that some air-traffic controllers are quite happy in their jobs and love the stimulation, while others hate the job, are often absent from work, and suffer both physically and mentally. Physical condition is very important in minimizing allostatic load, as are our living and working environments. Finally, psychological health is critical to physical health. Many studies of cancer patients, for example, show that feeling helpless and hopeless can contribute to allostatic load, whereas a sense of optimism or a sense of control tends to minimize allostatic load.

Probably the most interesting and optimistic evidence for the view that there is some hope in the workplace is a study that took place a number of years ago in a Volvo factory in Sweden. The study involved workers in a traditional production line, which meant that people were doing a specific job over and over again. There was a lot of absenteeism, a lot of job dissatisfaction, and such physiological measurements as blood pressure showed elevated levels. Then the production lines were reorganized so that people worked in teams and everyone could do everyone else's job. The quality of the work improved immediately, as did the employees' attitude toward their job. Their health also improved: Their blood pressure went back to a normal range. So it is possible to influence physical health by changing the working environment.

Finally, what can each of us as individuals do about stress in our lives? On the physiological front, most physicians would say that probably the single most important thing is to engage in regular moderate exercise. We also need to regulate our dietary intake of fats, eliminate smoking, and moderate our consumption of alcohol, all of which have an impact on the production of stress hormones. Keeping regular hours and getting enough sleep is also important. The body produces ACTH and cortisol on a diurnal rhythm: Production should be up in the early morning, down in the late afternoon and evening. There is increasing evidence that good physical and mental health are associated with maintaining this diurnal rhythm, whereas disturbance of this rhythm is associated with such conditions as depression, with its correlation of high cortisol levels and hippocampal shrinkage. On the social and psychological fronts, we need to seek social support and minimize social isolation, which has been shown to increase allostatic load. In addition, we need to find things that interest and absorb us outside the workplace, a form of relaxation that can reduce stress. Finally, if all else fails and we are feeling hopeless and helpless, we should seek professional help.

All of these aspects of regulating stress suggest that our mental well-being and our physical well-being are inseparable, two halves of a whole. As scientists are discovering every day, anything that affects one will, sooner or later, affect the other.[4]

5

Emotions and Disease: A Balance of Molecules

Esther Sternberg

Physicians and their patients have long known, at least on an intuitive level, that emotional health seems to affect physical health and vice versa. The ancient physician Galen, for instance, observed that "melancholic" women were more likely than those of more cheerful, or "sanguine," temperament to develop breast cancer.[1] Doctors have also long observed that a patient's confidence in the physician and the prescribed treatment can influence whether the treatment is effective. Until relatively recently, however, the connection between emotions and disease—or between mind and body—could not be proved scientifically, in large part because many of the intricate workings of the body's immune system remained a mystery. No single organ of the body is in charge of immunity; instead, the immune system's components are like an army of many special forces, engaging the foe without specific orders from a central headquarters or commanding general.

> As Dr. Esther Sternberg, the chief of the section on neuroendocrine immunology and behavior at the National Institute of Mental Health, explains, researchers have now found that the brain's neuronal network and the body's immune system carry on a lively communication that also includes the endocrine system, hence the advent of the relatively new discipline of neuroendocrine immunology. The three-way conversation occurs at the level of chemical molecules, involving neurotransmitters produced by the brain, hormones produced by the endocrine system, and specialized chemicals called cytokines produced by immune cells. Researchers are beginning to suspect that when this molecular signaling system gets out of whack, the result can be anything from inflammatory diseases such as rheumatoid arthritis to mood disorders such as depression.[2]

In "EMOTIONS AND DISEASE," an exhibition at the National Library of Medicine, there is a woodcut from a sixteenth-century Venetian medical textbook. In it, the physician Galen is shown diagnosing a lovesick young woman. "I came to the conclusion that she was suffering from one of two things," Galen is depicted as saying. "Either from a melancholy dependent on black bile, or else from trouble about something she was unwilling to confess." In the twelfth century, the noted Torah scholar, philosopher, and physician Rabbi Moses ben Maimon (Maimonides) made a similar connection between mind and body when he asserted that "the physician should make every effort that all the sick and all the healthy should be most cheerful of soul at all times" and, further, that "they should be relieved of the passions of the psyche that cause anxiety."

The concept that emotions and disease are linked has been around for thousands of years.[3] Indeed, in every era, Western

Figure 8 A medieval woodcut portrays the effects of the four bodily humors that ancient physicians believed acted on the human psyche. A man with an excess of black bile is too melancholy to get out of bed (upper left), while a surfeit of blood turns a sanguine man into a romantic, wooing his lady with music (upper right). A choleric man driven by too much yellow bile beats his wife (lower left), and phlegm makes a woman an unresponsive mistress (lower right). Adapted by Leigh Coriale Design and Illustration, with permission of Zentralbibliothek Zurik.

culture has tried to explain the apparent link using the best available tools. Galen believed that disease resulted from imbalances in the body's stores of black bile, yellow bile, blood, and phlegm [Figure 8]. Maimonides spoke of "passions of the psyche," or soul, as interfering with health. By the nineteenth century, following the rise of the great anatomists in the sixteenth century, physicians had come to believe that all diseases resulted from abnormalities of the anatomy.

However, there was a category of disease that frustrated physicians and scientists alike, because no anatomical abnormalities could be found. The classic example in Sigmund Freud's time was the so-called hysterical woman, who was prone to pains and illnesses that had no apparent underlying physical cause. The male equivalent of female hysteria was the breakdown that some soldiers suffered in battle, known in World War I as shell shock. (In World War II it was called war neurosis or battle fatigue; since the Vietnam War it has been known as post-traumatic stress disorder.) These illnesses were called functional neuroses, reflecting the notion that if there was no abnormality of anatomy, then the person was simply malingering or was a hypochondriac. Freud, along with others in this era, developed the theories of psychoanalysis to try to explain these mysterious illnesses, but they remained elusive.

In the 1920s and '30s, the American psychologist Helen Flanders Dunbar took the idea of hysteria or war neurosis a step further. Dunbar and some of her colleagues, including Franz Alexander, tried to explain many physical illnesses such as asthma, gastric ulcer, and heart disease in terms of abnormalities of the psyche. Dunbar thus coined the phrase "psychosomatic medicine" to refer to efforts to treat somatic, or bodily, illnesses arising from psychic disturbance. If nineteenth-century physicians wanted to blame all illness on anatomy, some in the early twentieth century were attempting to explain illness in terms of the patient's psychic and emotional history.

Science was not up to the task of supporting these efforts, however. There simply wasn't enough known about the biochemical, physiological, and hormonal underpinnings of the relationships between body and psyche to make a strong case for these theories. As a result, the idea that emotions and disease were linked again fell into disrepute among more traditional scientists and physicians, who demanded proof before they would accept this connection.

As physicians concentrated on purely physical remedies, the late 1800s and early 1900s saw the advent of pharmaceuticals and patent medicines. Many patent medicines contained large amounts of alcohol, some contained opiates, and all were touted to be useful for practically any ill from fevers to worms to teething, diarrhea, dysentery, nervousness, prostration, and kidney disease. Academic physicians of the day were concerned that the general populace was too gullible, so easily taken in people were by so-called cures that had not been proved. Eventually, though, physicians recognized that there was, in fact, something to be said for these unproved medications. In some instances, patients seemed to experience a cure, or at least a relief of symptoms, despite the fact that the substance they had taken involved no pharmacological action.

This phenomenon came to be known as the placebo effect—an amelioration of symptoms that seemed to be the result simply of the patient's belief in the medicine. A placebo (from the Latin for "I shall be pleasing")[4] was often just a sugar pill, variously colored to reflect the illness the doctor was ostensibly treating; indeed, as recently as a few decades ago many family doctors often stocked pills labeled Obecalp (placebo spelled backward).[5] Today, the development and testing of new drugs always involves administering a placebo as a control, in order to determine what component of a treatment is a biological effect of the drug and what component stems from this placebo effect.

Finally, in the 1950s, as researchers gained a better understanding of the physiology of the brain, they began to establish a link between emotions and disease in terms of the brain's responses to stimuli from the environment. The mind-body relationships were examined largely within the context of the subcortical brain—the autopilot part of the brain that controls our nonconscious, automatic responses, especially to perceived threat. The physiologist Hans Selye extended this idea when he proposed that the body's response to some sort of external stimulation or stress could cause disease.

Since then, technological advances have increased our understanding of this connection between stress and disease. Such instruments as PET scanners have given us phenomenal insight into the workings of the brain. For example, PET scans show that when we silently read a word, a certain part of the brain becomes activated, with increased blood flow and metabolism; another part becomes active when we think about the meaning of a word; and yet another part lights up when we say a word. [Figure 9] Similarly, we can observe the parts of the brain that become more or less active during different mood states, such as transient happiness or transient sadness. With the advent of molecular biology and computerized microscopes, we can even detect the proteins produced by genes that are activated in different parts of the brain under different conditions. For example, cells in the hypothalamus of a rat that has been exposed to some form of stress become activated and produce many different molecules, in-

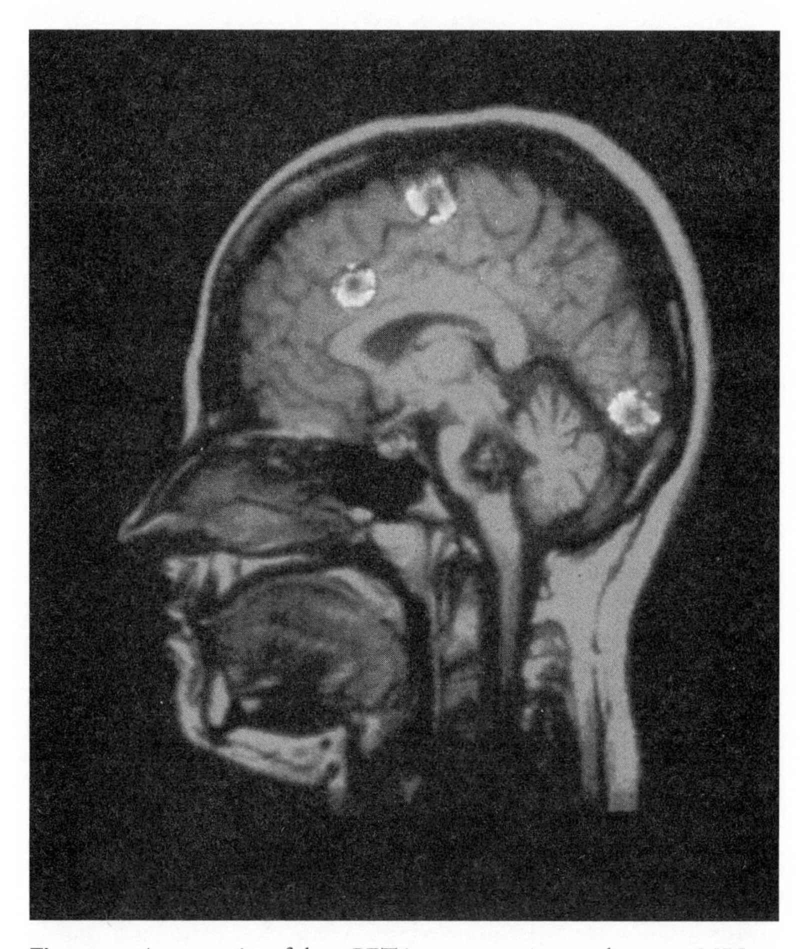

Figure 9 A composite of three PET images superimposed over an MRI scan of neural structure shows three distinct brain activities in the left hemisphere. Speaking a word activates a region in the brain's motor cortex, at the top of the image; silent reading engages an area in the visual cortex at the lower right, at the back of the brain; and thinking about a word's meaning triggers activity in the frontal lobe. Image courtesy of Dr. Marcus Raichle, Washington University, St. Louis, MO.

cluding the stress hormone CRH, or corticotropin-releasing hormone.

Cross Talk

Until fairly recently, researchers believed the body's immune defenses were independent of the central nervous system (CNS), or the endocrine system.[6] The troops of the immune system, white blood cells, for example, are generated and ferried about by the lymphatic system, a network of vessels and immune organs such as the bone marrow, the spleen, and the thymus. White blood cells, also called leukocytes, circulate both in the blood and in the lymph, constantly patrolling for foreign agents, or antigens, such as viruses, bacteria, dust particles, and pollen. Other immune cells congregate in lymphoid organs such as the lymph nodes, the spleen, and the appendix, where they work together to combat foreign organisms.[7]

In the past fifteen years or so,[8] however, researchers have discovered that there are not only direct physical links between the brain and the immune system (nerves that signal immune organs such as the lymph nodes and the spleen) but that there are also complex molecular connections between the brain, the immune system, and the endocrine system. Neurotransmitters, it turns out, are not confined just to transmitting signals between neurons; nor are hormones limited to the endocrine system. And immune-system chemicals, in turn, are not confined to the immune system, but also affect the brain.[9]

Much of this interaction occurs through the activation of the hypothalamic-pituitary-adrenal (HPA) axis in response to some form of stress, whether it's the sight of a predator, a near-accident on the highway, or, it turns out, injury or infection.

Since the hormones of this axis, corticosteroids, have profound effects on immune responses, we have begun to home in on how stress can influence the onset and course of disease.

To understand how this comes about, we need to look briefly at how the CNS and the immune system operate. Both the CNS and the immune system receive information from the environment and other parts of the body through various sensory elements. In the case of the brain, input comes from such senses as the eyes and the ears. In the case of the immune system, the sensory elements are the myriad patrolling immune cells whose duty it is to distinguish self from nonself—that is, to recognize what belongs in the body and what doesn't. Both systems also carry out an appropriate response through the action of various motor elements. Depending on what the eyes or the ears are relaying to the brain, an example of a motor response might be to turn away or cover the ears. The immune system's response to a foreign invasion is to launch a counterattack by scavenger macrophages—white blood cells that literally gobble up foreign matter—or, if that response is insufficient, to rally more specialized immune troops, resulting in activity that can result in inflammation and swelling at the injured or infected site.

The secret to how the brain-immune system link affects the onset and outcome of disease is that both systems use chemical mediators, specific messenger molecules, for communication. Further, these messenger molecules, whether they are produced by neurons in the brain or by immune cells calling for backup, can act as signals not only within their own systems but between the two systems as well.

The immune cell messengers, as it happens, constitute one of the major discoveries of contemporary immunology. Scientists have learned that white blood cells produce small proteins, known as cytokines, that indirectly coordinate the responses of other parts of the immune system to pathogens in order to fight infection, kill viruses, or trigger allergic reactions. Cytokines such as interleukin-1 and interleukin-2 also act as chemical signals between immune cells and other types of cells and organs, including parts of the brain. Researchers used to think that the so-called blood-brain barrier—the overlapping, tightly sealed arrangement of the cells that form the brain's fine network of capillaries[10]—guards the central nervous system against all potentially harmful molecules in the bloodstream. However, during inflammation or illness—and, to some degree, even in health—cytokines can cross the barrier at permeable sites and may be carried over into the brain, where they recruit the brain in the immune system's battle by triggering the HPA axis. Some research also indicates that cytokines might act more quickly to signal the brain, without having to travel all the way to the brain itself, by infiltrating clumps of nerves that in turn stimulate the vagus nerve, one of the cranial nerves that travel from the brain stem to the chest and the abdomen.[11]

One of the brain's responses to these signals from the immune system is fever. The hypothalamus in effect raises the body's temperature setpoint and then directs a number of body-warming activities, including shivering, which generates heat by getting the muscles moving, and increasing metabolism. The exact function of fever is still somewhat mysterious, but the resulting elevated temperature may destroy some bacterial or viral invaders, and it may also activate cells that produce infection-fighting antibodies.

Another of the brain's responses is so-called sickness behavior. When you have the flu, for example, you feel tired, achy, and feverish; you don't want to eat, you don't want to move. All of these behaviors, which are nature's way of persuading the injured organism to keep still in order to conserve energy for the battle against the illness or injury, are caused by the effects of cytokines like interleukin-1. These immune signal molecules can bind to receptors in the cells that line the wall of blood vessels and stimulate intermediary molecules, called prostaglandins, which then go on to stimulate neurons in the surrounding tissue, producing the achy feeling or fever, for example. One of the ways in which drugs like aspirin help to relieve flu symptoms is by suppressing the action of prostaglandins and blocking these sickness behaviors.

Another way in which cytokines like interleukins affect the brain is by being toxic to neurons themselves. When too much cytokine is expressed within the brain, nerve cells begin to die. Overexpression of interleukin within the brains of patients who have AIDS may explain some of the brain-damaging features of this syndrome and some of the dementia that goes along with it. This may also be part of the mechanism of neuron death in Alzheimer's disease. There is now evidence that increased inflammation within brain tissue is a component of Alzheimer's. Such inflammation and the resultant release of cytokines can damage neurons and contribute to dementia.

The Unifying Messenger

The chemical messenger that effectively unites the stress and immune responses is CRH, or corticotropin releasing hormone, the endocrine hormone released by the hypothalamus as the first step in the chemical cascade of the brain's response to stress. [Figure 10] One of the chief products of this chemical cascade, of course, is cortisol, which helps in the body's readiness to face danger by, among other things, increasing levels of blood sugar. This agent is not only the best-known hormone of the stress

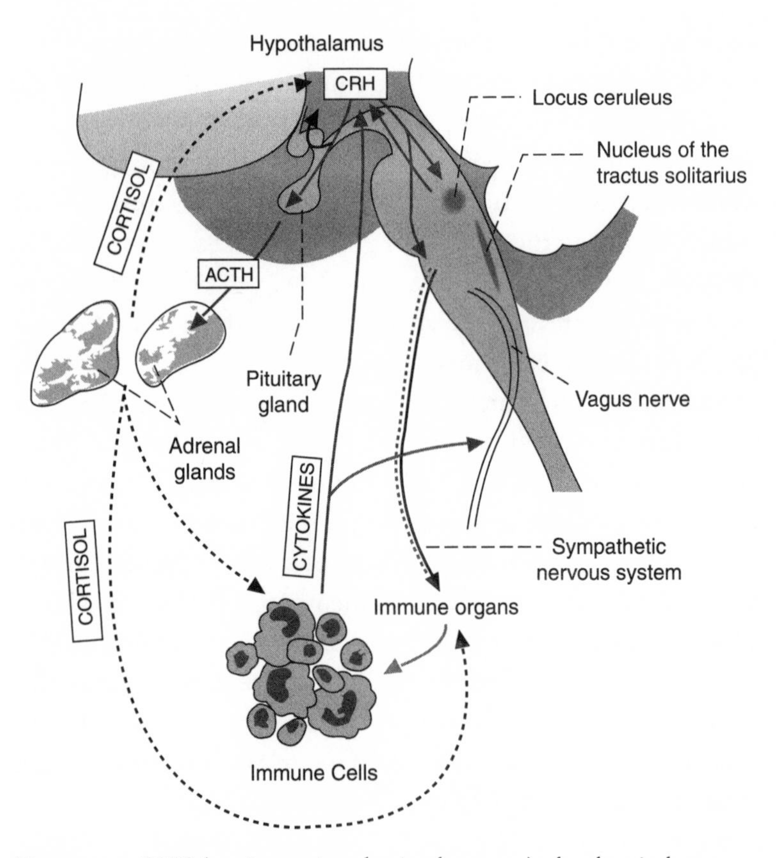

Figure 10 CRH (corticotropin-releasing hormone), the chemical messenger released by the hypothalamus during the stress response, is a crucial link between the brain and the immune system. As illustrated here, mediators from the immune and nervous systems can either stimulate (solid arrows) or inhibit (dotted arrows) their target organs, and vice versa. Chemicals called cytokines produced by immune cells stimulate the hypothalamus to produce CRH, thus activating the HPA axis and ultimately triggering the production of cortisol. Cortisol, in turn, suppresses inflammatory responses. If the intricate feedback loop malfunctions, the immune system can turn against the body itself, or become so underactive as to increase vulnerability to infection. From *Scientific American,* special issue, 1997; adapted by Leigh Coriale Design and Illustration, with permission of Roberto Osti.

response, it is also a potent anti-inflammatory and immunoregulator; by inhibiting the production of interleukin-1,[12] cortisol prevents the immune system from overreacting to injury and harming healthy cells and tissue. Once in circulation, cortisol inhibits the release of CRH by the hypothalamus, a simple feedback loop that signals the hypothalamus to shut down this production. If this feedback loop malfunctions, an oversupply or an undersupply of cortisol can have disastrous effects, resulting in an underactive or hyperactive immune system.

In addition to triggering the production of cortisol, the CRH-secreting neurons of the hypothalamus also reach to regions in the brain stem that regulate the sympathetic nervous system. As well as mobilizing the body during the stress response, the sympathetic nervous system innervates such immune organs as the thymus, the lymph nodes, and the spleen and thus helps to control the body's inflammatory responses. The hypothalamic neurons that produce CRH also reach to another brainstem structure called the locus ceruleus (the "blue place," for the color of its cells) and to the amygdala, which work together to control fear-related behaviors such as arousal, avoidance, and enhanced vigilance—behavior that can be useful in recovery from illness or injury. In this way, CRH and cortisol link the body's immune response and its brain-regulated stress response.

Keeping the System in Working Order

Given the many interactions and communications that take place between the central nervous system and the immune system at the molecular level, it is clear that maintaining good communication between the brain and the immune system is essential to health. As human beings moving about in the world, we're constantly exposed to many chemical and physical stimuli, including bacteria, allergens, and viruses. The immune system must not only be able to turn on at a moment's notice to fight these foreign invaders, it must also be able to shut off at a moment's notice when the foreign invader is gone, lest its continued activity damage the body itself.

Any interruption of the HPA axis—whether because of an inherited disorder or through the interference of drugs or surgery—that results in an undersupply of cortisol can in turn result in a hyperactive immune system and susceptibility to inflammatory diseases like arthritis. On the other hand, too much stimulation of the HPA axis—such as from chronic stress—will cause an overabundance of cortisol to be released. Excessive cortisol will shut off the immune system before it is able to clear the foreign invader, a situation that can result in increased susceptibility to infection.

Much research on immune-system response has been done using an inbred strain of laboratory rat called the Lewis rat. Lewis rats, when kept in a protected, clean environment, live long and healthy lives. But when they come in contact with a variety of common pathogens and antigens in the environment, they develop a host of inflammatory diseases, including arthritis, thyroiditis, adrenalitis, and uveitis (inflammation of the eye). Under certain conditions, they can also develop a disease that's very much like multiple sclerosis.

Surprisingly, the reason Lewis rats develop so many inflammatory illnesses is not, as one might think, because of a problem in their immune system. Rather, it is a problem in the central nervous system—specifically, in the hypothalamus, the structure that releases CRH and controls the stress response. This becomes evident by comparison with another strain of rat called

Fischer rats. This strain is histocompatible with Lewis rats, which means that the two strains can accept each other's grafts. However, Fischer rats develop relatively little inflammatory disease when exposed to the same kinds of environmental triggers that induce inflammation in Lewis rats. The difference between the two strains of rat is in the hypothalamus—in how much CRH each strain makes in response to a variety of stimuli. When exposed to the immune signal interleukin-1, for example, which would normally trigger the hypothalamus to secrete CRH, Lewis rats make very little, if any, of the hormone, whereas Fischer rats make a lot.

But further proof that the Lewis rat's problem is in the hypothalamus comes from experiments in which brain tissue from the disease-resistant Fischer rat is transplanted into the brain of the susceptible Lewis rat. If a Lewis rat is injected with a nontoxic foreign substance such as carrageenan, a type of algae used as a food stabilizer, the rat will develop a large inflammatory pouch. But a Lewis rat that has received a hypothalamic tissue transplant from a Fischer rat will not develop inflammation. These findings suggest new avenues of investigation for developing treatments for inflammatory diseases; such treatments might be made possible by developing drugs that target the parts of the brain that control CRH and the stress response rather than simply targeting the immune system.

Emotions and Disease

Although most of what we know about the relationship of the brain's stress-response system and the immune system comes from studies done on animals, these principles are increasingly being proved in human disease. There is growing evidence, for example, that a wide variety of human diseases are associated with impairment of the HPA axis and reduced levels of CRH. Lower levels of CRH mean lower levels of cortisol, which in turn result in a hyperactive immune system. In humans, impaired CRH function is associated with lethargy, fatigue, increased sleep, muscle and joint aches—classic symptoms of such illnesses as chronic fatigue syndrome, fibromyalgia, and rheumatoid arthritis, as well as of such mood disorders as seasonal affective disorder (SAD) and a form of depression called atypical depression, which is characterized by profound fatigue, increased sleep, and increased food intake.

The opposite condition—too much CRH—is often associated with melancholia, the classic form of depression. Melancholic patients typically suffer from, among other things, insomnia and an inhibition of eating, sexual activity, and menstruation. These reactions are characteristic of what happens when the body is physiologically aroused as a result of very high concentrations of CRH, the classic fight-or-flight hormone. In hunted animals, for instance, the sight of a predator elicits a typical pattern of behavior brought on by CRH: The prey freezes, focuses all attention on the predator, and ceases to either feed or reproduce. People who suffer from depression show the equivalent behavioral responses in the absence of apparent external stimulus, and the responses are carried to the extreme—behavior that becomes deleterious rather than life-saving.

Doctors have long recognized that there is a relationship between depression and inflammatory diseases such as arthritis without knowing how the connection might arise. It now seems likely that the relationship can be understood in terms of the underlying biology and pathophysiology of both these illnesses. Many of the symptoms of the two illnesses are the same: difficulty getting going in the morning, loss of energy, fatigue, apathy. It may be that these are related not simply because the pain of chronic arthritis causes depression, and not simply because severe stress may in some way precipitate arthritis, but perhaps because both of these illnesses are linked by an underlying problem in the HPA axis (the part of the brain that controls the stress response), so that too little CRH is released when necessary in both conditions. The nineteenth-century French Impressionist painter Pierre-Auguste Renoir, for example, is a case of someone who had severe rheumatoid arthritis and also suffered from depression. Renoir was known to paint some of his beautiful, light, and airy paintings even during flare-ups of arthritis. But when he was depressed he either painted more somber pictures or he lost the drive to paint entirely.

It may be that other arthritis patients who suffer from depression have a hormonal impairment that underlies both inflammatory disease and depression, which can perhaps lead to an episode of either illness, depending on whether the external stimulus is a physical or a psychological stress. Indeed, depression in arthritic patients does not always coincide with flare-ups of their arthritis. Such findings suggest that the classification of illnesses as either medical or psychiatric may be an artificial distinction. Psychoactive drugs may in some instances be used to treat inflammatory diseases, for example, and drugs that operate in the immune system might be useful for certain psychiatric disorders.

Finally, we know from animal studies that stress can affect the course and severity of such illnesses as influenza and infections with other viruses, including HIV. Similarly, human studies have shown that the flu vaccine is less effective in people who are caring for spouses with Alzheimer's disease—a notably stressful role—than in people of similar age and background who are not caretakers.[13]

As we've seen, the new understanding of the interaction between the stress and the immune systems explains how this might come about; it also explains the studies that suggest that classical psychological conditioning of animals could affect their immune responses. Classical psychological conditioning was first demonstrated by the Russian physiologist Ivan Pavlov when he taught dogs to salivate at the sound of a bell, a stimulus that is not usually related to eating. That is, an animal is taught to associate two unrelated stimuli (in Pavlov's case, a bell and some morsels of food). The animal has an instinctive response to one of the stimuli (in this case salivating at the sight and smell of food) but is eventually conditioned to have the same response when presented with the other stimulus alone.

In the mid-1970s, Robert Ader and Nicholas Cohen of the University of Rochester, while working with rats, trained the animals to dislike saccharin-flavored water by pairing it with a drug that made them ill.[14] The drug Ader had chosen, an anticancer drug, also suppressed the rats' immune systems, halting the normal division of white blood cells.[15] After enough trials, the saccharin-water alone produced a decrease in the rats' immune response similar to that produced by the drug.

What implications does such psychological conditioning have for humans?

The fact that the stress and the immune systems interact and have certain hormonal responses in common suggests that conscious attempts to modify one's reactions and responsiveness to stress could be effective.[16] Of course, investigators still don't know what proportion of an individual's response to various kinds of stress is genetically determined and how much may be consciously controlled through such practices as meditation and relaxation therapy. Clearly, animal research and research in humans, such as that carried out by Jerome Kagan and others, is teaching us that we each have our own stress-responsiveness level. A combination of genes and early experience can shape the brain's wiring in such a way that some people's stress thermostats are on high and others' thermostats are set on low. A comparison of the two strains of rat mentioned here, for instance, shows that the Lewis rats, whose hypothalamus does not produce much CRH, remain calm in novel situations. By contrast, the Fischer rats, which produce a lot of CRH, are very twitchy all the time.

Just as studies show that stress tends to blunt the body's immune responses, making us more susceptible to infection and disease, research also shows that a supportive social environment or group therapy, by reducing stress hormone levels, can enhance immune response, including resistance to such diseases as cancer. For instance, women with breast cancer who have a strong supportive network of friends and family tend to live longer than their counterparts who lack such support. Recognizing our individual styles of handling stress—whether we're prone to just feel anxious, for example, or talk to a friend, or head out the door for a run—can help us modify our stress responses if necessary, and can help maintain optimal activity of the immune system and a balance of health.

6

The Power of Emotions

Joseph LeDoux

Whether we're seven or seventeen or seventy, emotions can often be very confusing. For all that has been written and sung and otherwise celebrated about love, or anger, or grief, in some ways it's as if we all spoke a different language when it comes to explaining what we feel. In large part because emotion is such a subjective experience, scientists have long been wary of trying to study the brain's role in it, preferring to concentrate on more easily measured mental processes such as perception and memory.

At the same time, though, human beings connect with one another on an emotional level. We react in similar ways to certain stimuli. A theater full of individuals can be brought to tears by one movie scene, to laughter by another. If we witnessed a fatal car accident a few weeks ago, our hearts might pound at the sound of screeching brakes today.

Pioneering neuroscientists like Dr. Joseph LeDoux, the Henry and Lucy Moses Professor of Science at New York University Center for Neuroscience, have begun to examine the way the brain shapes our experience—and our memories—to generate the varied repertoire of human emotions. Specifically, as LeDoux explains in this chapter, he chose to begin the inquiry by examining an emotion that is common to all living creatures: fear. In unraveling the workings of the brain's mechanisms for detecting and responding to danger—and remembering what is dangerous—LeDoux has found that this system involves neural pathways that bypass the higher, "thinking" parts of the brain. In so doing, the fear system creates what LeDoux calls emotional memories (as distinguished from memories *about* emotion)—memories that trigger the physiological responses that can wash over us without our having the least idea why.

"The only questions worth asking are whether humans will have emotions tomorrow and what life would be like if the answer is No."

—rock critic Lester Bangs

UNDERSTANDING emotions ought to be an important part of any science that's concerned with how the brain and the mind work. Our emotions make up who we are, after all; they color what we want to be—and what we don't want to be. Unfortunately, many recent neuroscientific approaches to studying the mind have left emotions out. Emotions have mostly been studied psychologically in modern times. Such efforts have provided insights, to be sure, but they also have a couple of drawbacks. One is that, in effect, everybody knows what emotion is, but no one seems to be able to define it. The other is that there are as many theories of emotion as there are workers in the field. But studies of the brain can provide new insights into how a psychological process like emotion might work and are a valuable approach.

Listening to the Brain

For many years, some people felt that the problem of how the brain "does" emotion had been more or less solved. In the 1950s, Paul McLean, a scientist in the Laboratory of Neurophysiology at the National Institute of Mental Health, published a paper theorizing that the limbic system (a designation that includes the hippocampus, the thalamus, the hypothalamus, and the amygdala, among other structures) could be regarded as the seat of our emotions.

The problem was that the limbic-system theory tried to be a theory of all emotions at once. However, the brain does not have a single system dedicated to the function of "emotion." Instead, the various classes of the mental states that we collectively refer to as emotion are mediated by separate neural systems, each of which evolved for different reasons. The system that evolved to defend against danger, for example, is likely to be different from the one that evolved for procreation, so the subjective feelings that occur when these systems are activated—feelings of fear and of sexual pleasure—have unique neural underpinnings, at least in part.

The brain mechanisms that generate a given mental state, or what we choose, for the sake of convenience to call emotion, also give rise to certain measurable physiological states, such as pulse rates or brain waves, as well as to observable behaviors such as running away or smiling. "Feelings," by contrast, are a conscious, subjective experience. A person can say "I feel afraid," but that assessment depends on the individual's own labeling of his or her state. Another person, displaying some of the same physiological responses (a rapid pulse, say, or sweaty palms) might say "I feel excited." So trying to work backward from our feelings into the brain is fraught with problems. What we have to do instead is build up our understanding of what we call emotions by looking at the systems that evolution designed to produce them. Consequently, my approach has been to let the brain tell me how individual emotions are represented in it—what the brain mechanisms are, in other words.

Fear as a Model System

The fear system is a useful place to begin for a variety of reasons. First, fear is pervasive. We humans may not have to be afraid of being attacked by bloodthirsty predators on a daily basis, but modern society presents us with countless other dangers—from potential worldwide catastrophes such as nuclear war to more commonplace personal events such as being the victim of a street crime.

Another reason fear is a good emotion to study is that it's at the root of many psychiatric problems. The so-called anxiety disorders—panic attacks, obsessive-compulsive disorder, post-traumatic stress disorder—make up about half of all the psychiatric conditions that are treated every year, not including substance-abuse problems. Because these are all disorders of the brain's fear system, an understanding of this system could have important clinical implications.

Finally, the brain system that generates fear behavior evolved to help animals stay alive and has been preserved for millions of years, across a variety of species. The way that we act when we're afraid—the way the body responds—is very similar to the way that other animals act when they're afraid, even though we aren't reacting to the same things. A rat would never be sent into a panic attack by the news that the stock market had crashed, and a human is not, ordinarily, afraid of a cat. But the way our body responds to the news of a stock market crash is very similar to the way the rat's body responds when it sees a cat. This is critically important, because it means that we can study the behavior of other animals, and the processes in their brains, to learn how the human fear system works.

Behavioral Tools: Classical Fear Conditioning

So how do we go about the process of studying fear? Two things are necessary. First, we need good behavioral tools—that is, techniques and methods for studying such specific behavior as how an animal acts when it is afraid. And we also need good neuroscience tools, methods that allow us to study what is going on in the brain when the animal is behaving in a fearful way.

One important behavior tool is known as classical fear conditioning, which is a version of what Pavlov described as the conditioned reflex. The process of classical conditioning involves pairing, or associating, an innocuous stimulus—a sound or a flash of light, something that is essentially meaningless in itself—with something that is meaningful to the animal. In the case of Pavlov's dogs, the meaningful stimulus was food; the meaningless stimulus was the bell. Food is not a useful stimulus if we're interested in studying fear, however. So, using laboratory rats as subjects, we might pair a sound with, for instance, a mild foot shock. (We keep the shock as weak as possible to allow the experiments to be performed, and we administer it as infrequently as is feasible.) On the basis of these kinds of pairings, the sound becomes something that the rat learns is associated with danger. Thus when the rat hears the sound, it reacts immediately: It freezes in anticipation of danger. This is a conditioned reflex, as is Pavlov's dogs' salivating at the sound of the bell, in anticipation of the food.

An animal in the wild usually doesn't have the luxury of trial and error in learning what's dangerous; it doesn't get to practice until it gets things right. If it's lucky enough to escape once, it had better remember the sight of the predator, the smell of the predator, the sound of the predator, and so forth. In the laboratory, we need to apply the shock with the sound only once if it is sufficiently aversive. Typically, we use a weak shock and pair it with the sound several times to minimize discomfort.

When something like this occurs—the sound that's been paired with the shock—it activates a variety of responses that are identical to those that would occur in a real-life situation. Television tapes of the bombing during the 1996 Olympic Games in Atlanta, for example, reveal that when the bomb went off the first thing that happened was that everyone flinched; this was the startle reflex. But then the next thing they did was freeze: They just hunkered down and held still for about two seconds. That's evolution buying us a little time, left over from when we used to be at the mercy of various predators. Predators respond to movement; they can't necessarily see very well, but movement will attract them. So we freeze when we're in a dangerous situation, because our old evolutionary fear system detects danger and responds to it in an automatic way.

In addition to the startle and freeze reflexes, the body has other automatic reactions. These are all part of the so-called fight-or-flight response. In a situation of danger, a variety of physiological responses occur. Blood is redistributed to the body parts that are most in need (the muscles). This results in changes in blood pressure and heart rate. In addition, the hypothalamic-pituitary-adrenal, or HPA, axis is activated, releasing stress hormones. In general, the body is readied to move quickly. In addition, the brain activates the release of natural opiate peptides, morphinelike substances that block the sensation of pain.

Called hypoalgesia, this reaction is an evolutionary carryover that allows a wounded animal to keep going. It's often seen in wartime, where wounded soldiers don't react to their injuries until they're off the battlefield. All of these things happen in the rat when it perceives a natural threat such as a cat, or when it hears the sound that's been paired with the shock. And all of these fear responses—the freezing response, the changes in blood pressure, the pain suppression—are easily measured.

Neuroscience Tools

In addition to behavioral tools, we also need the tools of neuroscience to understand how the brain's fear system works. Three tools are generally used to figure out which parts of the brain are involved in a given activity or behavior. One is called a brain lesion, a small hole made in brain tissue to interrupt the flow of information between neurons. By blocking the flow of information in a given pathway with a lesion, we can determine whether that pathway is involved in the behavior we're studying. That is, lesions in some areas will have no effect on the behavior, and lesions in other areas will interfere with the behavior, thus implicating that area. People with strokes or tumors have natural lesions, which typically are not very precisely localized. In animal studies, though, we can go right to the area we need to examine. Considerable research has produced precise maps of the brain of the rat, and of many other animals as well. As a result, we can go into a specific region of the rat brain on the basis of three coordinates—left/right, up/down, and front/back—and make a lesion by releasing a small amount of current or injecting a chemical.

The brain maps are also useful when we want to measure the electrical activity of a particular region. Because communication between neurons is based on electrical activity, we can insert electrodes attached to amplifiers to record responses in a given area of the brain. When neurons are excited by input from another area, they generate signals called spikes or action potentials. Thus if neuron A activates neuron B, neuron B will fire these action potentials, which tells us that neuron B is part of the brain circuitry involved in the behavior we're studying. The advantage of this tool is that, unlike a lesion, it doesn't damage the brain in the process.

Finally, we can trace actual connections in the brain—determining whether Area X sends its axons to Area Y or to Area Z—by tracking chemical activity. To do tracing experiments, we inject a tracer substance into Area X using the road maps described above. The tracer is taken up by the neurons in the area injected, then hitches a ride on molecules that are being shipped down the axon. (Nutrients and neurotransmitter-related chemicals move back and forth within neurons all the time.) We can then stain or dye the brain to see where the substance appears next; the region will stain brightly enough so that we can see it under the microscope. This tells us which areas Area X talks to.

Once we have conditioned the animal to respond to a sound—so that the sound produces freezing behavior, changes in blood pressure, heart rate, and so forth—the next step is to trace how the sound, coming into the ear, reaches the parts of the brain that produce these responses in the body. The strategy is to make a lesion in a certain part of the brain to determine whether damage to that area interferes with the fear conditioning. If it does, we then inject the tracer substance there to see which areas that part of the brain communicates with. Then we systematically make lesions in each of those downstream areas to see which one interferes with the fear conditioning, inject tracer substance at that point, look to see where it goes, and so on. We can then record electrical activity to see how cells in the area respond. In this way, we can walk our way, point by point, through whatever pathway of the brain we want to study and make some sense of how it works.

Brain Pathways in Fear Learning

Years of research by many workers have given us extensive knowledge of the neural pathways involved in processing acoustic information, which is an excellent starting point for examining the neurological foundations of fear. The natural flow of auditory information—the way you hear music, speech, or anything else—is that the sound comes into the ear, enters the brain, goes up to a region called the auditory midbrain, then to the auditory thalamus, and ultimately to the auditory cortex. Thus, in the auditory pathway, as in other sensory systems, the cortex is the highest level of processing.

So the first question we asked when we began these studies of the fear system was: Does the sound have to go all the way to the auditory cortex in order for the rat to learn that the sound paired with the shock is dangerous? When we made lesions in the auditory cortex, we found that the animal could still make the association between the sound and the shock, and would still react with fear behavior to the sound alone. Since information from all our senses is processed in the cortex—which ultimately allows us to become conscious of seeing the predator or hearing the sound—the fact that the cortex didn't seem to be necessary to fear conditioning was both intriguing and mystifying. We wanted to understand how something as important as the emotion of fear could be mediated by the brain if it wasn't going into the cortex, where all the higher processes occur. So we next made lesions in the auditory thalamus and then in the auditory midbrain. The midbrain supplies the major sensory input to the thalamus, which in turn supplies the major sensory input to the cortex. What we found was that lesions in either of these subcortical areas completely eliminated the rat's susceptibility to fear conditioning. If the lesions were made in an unconditioned rat, the animal could not learn to make the association between sound and shock, and if the lesions were made on a rat that had already been conditioned to fear the sound, it would no longer react to the sound.

But if the stimulus didn't have to reach the cortex, where was it going from the thalamus? Some other area or areas of the brain must receive information from the thalamus and establish memories about experiences that stimulate a fear response. To find out, we made a tracer injection in the auditory thalamus (the part of the thalamus that processes sounds) and found that some cells in this structure projected axons into the amygdala. This is key, because the amygdala has for many years been known to be important in emotional responses. So it appeared that information went to the amygdala from the thalamus without going to the neocortex.

We then did experiments with rats that had amygdala lesions, measuring freezing and blood-pressure responses elicited by the sound after conditioning. We found that the amygdala lesion prevented conditioning from taking place. In fact, the responses are very similar to those of unconditioned animals that hear the sound for the first time, without getting the shock.

So the amygdala is critical to this pathway. It receives information about the outside world directly from the thalamus, and immediately sets in motion a variety of bodily responses. We call this thalamo-amygdala pathway the low road because it's not taking advantage of all of the higher-level information processing that occurs in the neocortex, which also communicates with the amygdala. [Figure 11]

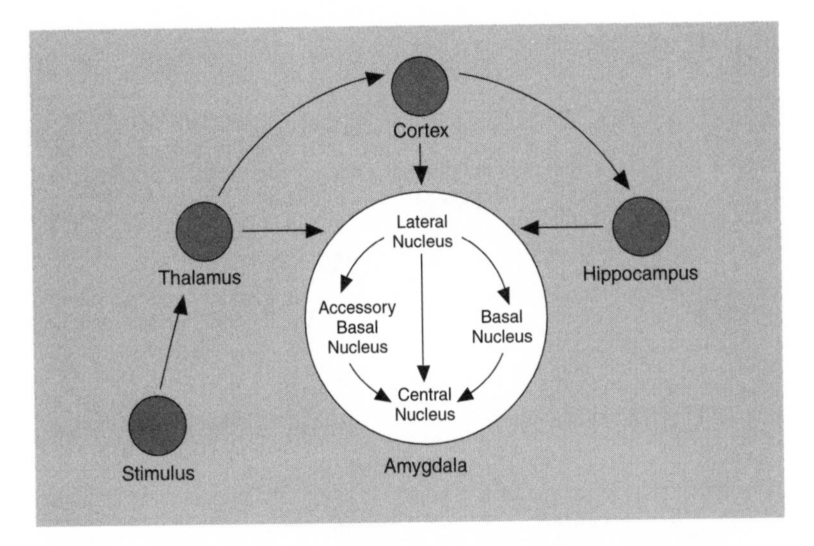

Figure 11 Different parts of the amygdala receive different sensory inputs but they all communicate with the central nucleus, which communicates with the brain stem, thus setting in motion physiological responses. A key player in our emotional behavior, the amygdala also receives sensory information directly from the thalamus—the so-called "low road"—and triggers a variety of bodily responses even before the information has been processed by the cortex. Adapted by Leigh Coriale Design and Illustration from *Emotion, Memory and the Brain,* by Joseph E. LeDoux. © 1994 by Scientific American, Inc. All rights reserved. Used with permission.

Unconscious Emotional Reactions

As an example of how we think these pathways work, let's say that a hiker is walking through the woods and sees something on the ground. [Figure 12] The image gets to the thalamus, which sends a very crude template to the amygdala; the amygdala, in turn, activates the heart rate, gets the muscles tense and ready to go. At the same time, the stimulus

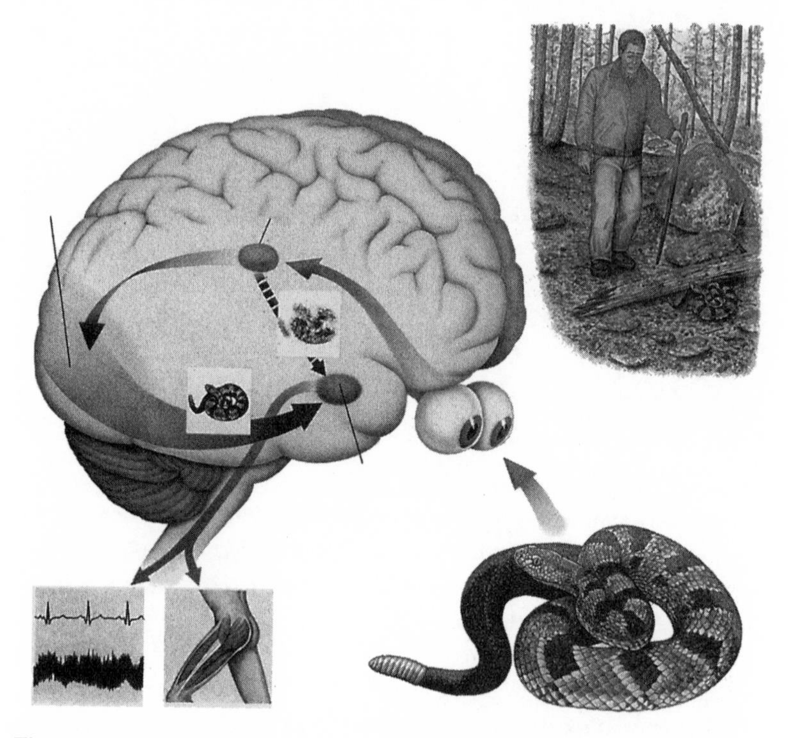

Figure 12 One example of how the thalamo-amygdala pathway works is depicted here, with a hiker in the woods who sees a shape on the ground. The visual stimuli are processed by the thalamus, which passes a crude template of the shape directly to the amygdala. The brain starts responding instantly, raising heart rate and blood pressure, and causing muscles to contract, in preparation for fight or flight. When the information from the thalamus reaches the visual cortex, the brain can determine whether the shape is actually a snake or not—in which case, the fear response is allowed to go forward or is quelled. Illustration by Robert Osti. Used with permission.

is making its way through the cortex, which is slowly building up a complete representation of—a snake. Now, the thalamus doesn't know if it's a snake or a just a stick that looks like a snake, but as far as the amygdala is concerned in this situation you're better off treating the stick as a snake than you are treating a snake as a stick. The subcortical brain is overgeneralizing for the opportunity to stay alive in the presence of the snake. By getting the amygdala going instantly, it buys you time. If the object turns out to be a stick instead of a snake, nothing's lost; you can turn the fight-or-flight system off. But if it turns out to be a snake, you're ahead of the game: You've activated the amygdala, and your body is ready to respond effectively.

The low road, or the thalamo-amygdala pathway, is a quick and dirty system. Because it doesn't involve the cortex at all, it allows us to act first and think later. Or, rather, it lets evolution do the thinking for us, at least at the beginning, buying us time. We freeze first, and that gives us a few seconds to decide what to do: Run away? Hold still? Try to fight?

The cortex—the high road, so to speak—also processes the stimulus, but it takes a little longer. You need the cortex for high-level perception in order to distinguish one kind of music from another, for example, or to distinguish between two speech sounds. But you don't need the cortex to carry out some of the emotional learning involved in the fear system. Thus we can have emotional reactions to something without knowing what we're responding to—even as we start responding to it. In other words, we're dealing with the unconscious processing of emotion. This is a neurological demonstration of at least part of what Freud was trying to get at when he talked about unconscious emotions.

Without subscribing to the whole Freudian hypothesis at this point, what we're saying is that unconscious emotions are probably the rule rather than the exception. We all know that there are many times in normal, day-to-day experience when we don't understand where our emotions are coming from—why we feel happy, sad, afraid. For example, let's say you're in a restaurant having a meal with a friend and you have a terrible argument at the table, which happens to be covered with a red-and-white checkered tablecloth. The next day you're walking down the street and you have this gut feeling that the person walking toward you is someone you don't like. You've never seen the person before, but you know you don't like him. We often hear about "gut feelings" and people who say, "You have to trust your gut." But maybe in this case the reason you feel you don't like this person is simply that he's wearing a red-and-white checkered tie. This visual input is going in through your low road, activating your amygdala and causing you to have an unpleasant reaction to the person. You might attribute your reaction to the way this person looks or walks or acts, but in fact it's just the low road sending a cue to the amygdala and activating it unconsciously.

Some of the time, at least in evolutionary terms, these low-road reactions are useful. Certainly that was the evolutionary goal: to protect us from danger. But these can also be harmful, or at least counterproductive. As in the case of the red-and-white checkered tie/tablecloth, an unconscious response may not be revealing some sort of inner truth but may instead be doing nothing more than reviving past emotional learning. "Listening to our gut," in other words, might simply mean we are responding to past learning.

We can think of the amygdala as the hub in a wheel of fear, with some of the spokes being inputs and others outputs. We've already discussed the low-road input from the thalamus, but other areas of the brain provide input to the amygdala as well. Information about what we might call sensory objects—a visual object such as an apple, say, or complex sounds like music or speech—comes from the sensory cortex.

Other parts of the cortex are involved in higher cognition. For example, a cortical area called the hippocampus is involved in such higher-order aspects of cognition as long-term memory and the processing of the context of events, the kind of information that allows us to say where and when something happened, along with other elements of the scene, such as whether it was raining. If you damage or remove the hippocampus in rats, for instance, the animals are no longer able to recognize a familiar place; they are unable to distinguish whether the test chamber they're in is one where they've been conditioned to mild foot shocks. As a result, they express fear responses in all similar chambers. Let's say, for example, that you regard all snakes as dangerous, but you know that you needn't fear a snake in the zoo as much as you might a snake that you happened upon in the woods. Ordinarily, your hippocampus and cortex would recognize the context (are you in the woods or at the zoo?), and you would react appropriately to the sight of a snake. But if you had a hippocampal lesion, you might have trouble suppressing a strong fear reaction even at the zoo.

It's natural to think of the low road as unconscious and cortical processing as conscious. But, very likely, even information processed by the cortex remains unconscious until it becomes available to a special system in the frontal cortex, called working memory, as described below.

Getting Rid of Fear

Another important player in the fear response is the prefrontal cortex. In rat studies, as well as in human experiments, when you give the sound over and over again, without the unpleasant event occurring, it eventually loses its ability to elicit the emotional fear reactions. This process is called extinction. But if the medial part of the prefrontal cortex is damaged, emotional memory is difficult to extinguish. So, for example, a rat that has a lesion in the prefrontal cortex tends to continue to respond to the sound as if it were still associated with the unpleasant event; the learned response is resistant to extinction.

However, it's important to know that even without damage to the prefrontal cortex, fear memories are hard to extinguish completely. Many studies show, for example, that weeks after a rat has ceased to react to a sound that had been paired with a shock, it might suddenly react fearfully to the sound again. Or if the animal goes back into the chamber where it had the conditioning experience, the fear behavior can be reactivated. Stress can reactivate extinguished fears in humans as well. A patient with a phobia, for example, can be treated, apparently successfully; then something happens—say the patient's mother dies—and the phobia comes back. What certain types of therapy can do—and what the extinction process does—is train the prefrontal cortex to inhibit the output of the amygdala. This training doesn't eliminate the unconscious fear; it simply holds it in check.

Therapists find this both depressing and informative; they now understand that fear memories can't be completely eliminated, but at least they know what battle they're up against. Indeed, one of the problems in understanding human phobias has been figuring out how to study them in the laboratory. If an animal hears the sound over and over again without the shock, it stops responding fearfully relatively quickly. But taking someone who is deathly afraid of heights to the top of the

Empire State Building, without a lot of behavioral therapy beforehand, is likely to make the condition worse rather than better.

So the simple extinction process doesn't seem to work with phobias, leading some people to speculate that phobic learning is something special. It may be, however, that what's different is not the kind of learning that takes place in anxiety disorders like phobia but the kind of brain that's doing the learning. Just as we can create extinction-resistant learning in the rat by making a small lesion in the prefrontal cortex, it may be that something subtle is malfunctioning or changed in the prefrontal cortex of the phobic patient that makes him develop learning that is resistant to extinction.

Recent studies with Kevin LaBar and Liz Phelps at Yale, in which human subjects have been conditioned with sound and mild shock, show that the human brain works basically the same way as the rat brain—or, for that matter, the same way as the brains of primates, dogs, cats, rabbits, pigeons, and lizards. Damage to the human amygdala interferes with fear conditioning in people, and the human amygdala lights up during fear conditioning, as revealed by functional imaging. I don't know of any animal that can't be conditioned in this way, and in any animal that has an amygdala, that structure seems to be involved in fear conditioning. The fear system, therefore, is probably a very basic, fundamental learning mechanism that's built into the brain.

In this sense, then, we're emotional lizards. We're running around with an amygdala that's designed to detect danger and respond to it. Obviously, that's not all there is to an emotional reaction, but it's the way the emotional reactions get triggered. This system is very efficient, and it hasn't changed much in terms of how it works. What has changed, of course, are the kinds of things that will turn it on, the things that humans have learned to respond to that have the same effect on us that seeing a cat has on a rat.

Memories about Emotions

Of the several different memory systems in the brain, the hippocampus is involved in the system whose job is to create the kind of memories we usually mean when we say we "remember" something. You remember what you ate last night or where you went last week or what you did at your grandmother's when you were six years old. These are your memories, and they involve the hippocampus. Suppose you're driving down the road and you have an accident. You hit your head on the steering wheel and the horn gets stuck on. You're bleeding and you're in pain. The sound of the horn becomes linked in your brain with the pain, and this relation is stored in various brain systems. Days or weeks later, a horn goes off. The sound enters your brain and reminds you of where you were and who you were with and what you were doing the last time you heard it. It will also remind you that it as an awful experience, but that memory is a cold fact. None of these are emotional memories, which come through the amygdala system; instead, these are memories *about* the emotional experience. Emotional memory occurs when the sound reaches the amygdala, which activates your autonomic and hormone systems, and causes your muscles to tense up because of associations stored there. In other words, the hippocampal system gives you conscious memory of an emotional experience; the amygdala system gives you unconscious emotional memory.

Patients in whom the hippocampal system is damaged have poor conscious memory. For example, the neurological literature includes the famous case of a woman who had severe amnesia. Each day when the doctor walked into her room, he would have to reintroduce himself, because the woman never remembered having seen him the day before. In fact, if he left the room for just a few minutes, she wouldn't remember him when he returned. One day the doctor walked in and extended his hand to shake hers, something he did every time. But this time he held a pin in the palm of his hand. When their hands met, hers was pricked and she withdrew it immediately. The doctor then left the room, and when he came back a few minutes later he stuck out his hand again, as usual. This time she wouldn't shake hands with him. She had no conscious memory of being pricked by the doctor, nor did she have any conscious memory of the doctor, but unconsciously she was able to withdraw her hand and protect herself. The amygdala remembered.

By contrast, patients whose hippocampus is intact but who have amygdala damage are unable to do this kind of pinprick learning, this kind of fear conditioning. They know all the details—that the doctor was in the room, that they were pricked—but they don't withdraw their hand when the doctor tries to shake. After many decades in which neuroscientists thought there was only one kind of memory system, we now know that the amygdala and hippocampus systems mediate separate kinds of memory. Normally, they work together in such a way that emotional memories (mediated by the amygdala) and memories of emotion (mediated by the hippocampus) are fused in our conscious experience so immediately and so tightly that we cannot dissect them by introspection. Only by taking these systems apart in the brain, especially through studies of animal brains, have we learned about their separate existence.

The Effects of Trauma on Memory

A traumatic situation, in which an animal or a person is under stress, has separate consequences for these two kinds of memory systems. When the HPA axis releases stress hormones into the body, the hormones (especially cortisol) tend to inhibit the hippocampus, but they excite the amygdala. In other words, the amygdala will have no trouble forming emotional, unconscious memories of the event—and, in fact, will form even stronger memories because of the stress hormones. But the same hormones can interfere with the normal action of the hippocampus and prevent the formation of a conscious memory of the event.

This finding has a bearing on the Freudian concept of infantile amnesia—that is, the inability to remember things that happened to us before we were, say, about three years old. One possible explanation that scientists have proposed for infantile amnesia is that the hippocampus is not fully formed and functional until around age three. As a result, we're unable to develop long-term conscious memories of our life before that age. However, the amygdala is fully functional at a much earlier age. For this reason, children who are abused at a very early age might form strong emotional memories that they never have conscious access to for the rest of their lives.

This raises the question of whether anyone can ever recover "lost" memories. This is something that we can't deal with in terms of one individual. We can never say for sure that a given person has recovered a real lost memory or has what's called a false memory. Certainly considerable research has shown the fragility—and fallibility—of memories that seem highly detailed and realistic. However, unconscious emotional memories of trauma can affect people who have no conscious understanding of what is going on. The mere sight of the person or the instrument that was used in the trauma—a belt used in a beating, say—can activate the emotional system, causing panic attacks and fear responses that can also be generalized more broadly. Someone who was beaten with a belt as a very young child, for instance, might start out being afraid of only a belt, but the fear can generalize to leather, animals, and many other things. Unconscious emotional memories can therefore have widespread, long-lasting effects without our having any understanding of what is triggering certain responses or feelings.

Consciousness and Feelings

So where do "feelings" come in, then? I suggest that feelings enter the picture at the level of consciousness. Perhaps the best way to understand this idea is to look at the fear system again. We've seen that all animals have a fear-learning mechanism. The fear system evolved to produce adaptive responses, behavioral solutions to the problem of survival—how to detect danger and how to respond to it. The fruit fly can detect danger and respond to it, as can the bird, and so forth. But these animals don't necessarily have what we call fearful feelings the same way that human beings can "feel afraid" when they detect danger and respond to it. More important, from the standpoint of doing rigorous science, is that even if animals do have feelings, we have no way of knowing it. So we're better off assuming that they don't and then seeing how far we can proceed in the investigation of emotion systems in the brain without calling on feelings.

So we can say that the fear system evolved as a means of dealing with danger, by detecting it and producing evolution's best-guess responses about what to do in a dangerous situation—the freezing response, escape response, all the physiological preparation involved in the fight-or-flight response. Now, when a basic system such as a system for detecting danger arises in a brain that also has consciousness (that is, a brain that is aware of itself and of its relationship to the rest of the world), a new phenomenon occurs: subjective feelings. Feelings of fear, then, are what happen in consciousness when the activity generated by the subcortical neural system involved in detecting danger is perceived, in effect, by certain systems in the cortex, especially the working memory system (see below).

A conscious feeling of fearfulness is not necessary to trigger an emotional fear response. The low road can take care of this just fine. That is, we can produce responses to danger without being consciously afraid, as when we jump back up onto the curb to avoid being hit by a car that comes around the corner suddenly. In a situation like that, as people so often say, we don't "have time to be afraid." And at other times we will first have some kind of response in our body and only later be able to name what the feeling was ("anxious" or "angry" or "sad"). In many cases, though, even if we can say that we feel anxious, we don't know what generated those feelings. Indeed, we see this again and again in the various disorders of the fear system, such as panic attacks and phobias. People with panic disorder become terror-stricken for no apparent reason; someone with a phobia becomes irrationally afraid of heights or water, for instance, or even of leaving the house.

Why is it so difficult to eliminate such fears? Once the amygdala is turned on, it can influence information processing in the cortex from the earliest stages onward, but only the later stages of cortical processing affect the amygdala. In other words, even though communication goes two ways, it's not equally effective in both directions. In general, the projections from the amygdala to the cortex are much stronger than vice versa. If we think of the routes from the amygdala to the cortex as superhighways, then those from the cortex to the amygdala are narrow back roads. Once fears are activated, they can influence the entire working of the cortex, whereas the cortex is very inefficient at controlling the amygdala. So when therapists are working with people who have phobias, they have to use the back roads and side streets from the cortex to try to gain control of the amygdala—to interrupt the triggering of emotional memories—even as the amygdala is bombarding the cortex with input via the superhighways.

Psychotherapy might thus be viewed as, basically, a way to rewire your brain. If all changes in the brain are the result of learning experiences, then psychotherapy is a process of trying to teach the brain to unlearn some learned emotional associations. Learning—which is to say, rewiring the synaptic connections in the brain—requires that certain brain chemicals be squirted out at just the right place and in just the right quantities. In fact, as neuroscientists discover more about the molecular basis of all kinds of learning, it becomes clear that the molecules of memory are blind to the kind of memory—conscious or unconscious—that is occurring. What determines the peculiar quality of different kinds of memories is not the molecules that do the storing but, instead, the systems in which those molecules act. If they act in the hippocampal system, the memories recorded are factual and accessible to our consciousness. If they act in the amygdala system, they are emotional and largely inaccessible to our conscious awareness.

Emotional Working Memory

As we've seen, we often don't know what has triggered a given emotional response until after the fact. But at some point we do become aware of "feeling" a certain way. Where does this awareness occur? We have something called working memory, which is an important window into how our consciousness may be coming about. Working memory, which involves the frontal lobe, just above and behind the eyebrows, is what we use when we want to remember a new phone number long enough to dial it, for example, or to remember a decision we've just made long enough to carry it out. Let's say we're getting a glass of water in the kitchen and notice that the trash can is overflowing; working memory is where we hold the idea "I'll take out the trash" while we finish drinking the water. It is also a place where disparate kinds of information can be held simultaneously and compared to one another. For instance, we can have all kinds of information in working memory: the way something looks, sounds, and smells; along with, say, a memory of something that happened earlier in the day or a long time ago; and physiological input from the fear system (pounding heart, tense muscles). It's probably the only place in the brain where we can put all these elements together simultaneously. However, working memory can do only one thing at a time. That is, it can hold several pieces of information simultaneously, but only if they're all related to one job. The classic example is when we're trying to remember a phone number and someone asks us about something that's completely unrelated; the phone number very likely goes out the window. Since this is the way consciousness seems to work, many researchers are excited at the prospect that working memory may be, if not the basis of consciousness, at least a window into consciousness.

So working memory may be the system where conscious feelings occur. This is because three things come together in working memory: present stimuli, activation of the amygdala by those stimuli, and activation of conscious memory through the hippocampal system. Let's say, for example, that someone who is afraid of heights is standing at the top of the Empire State Building. Present stimuli might include the sight of people and cars on the street looking very, very tiny, and the whirring of a helicopter sounding very close. The activation of the amygdala by those stimuli would produce the unconscious arousal of emotion, the pounding heart and shortness of breath, and so forth. Activating conscious memory through the hippocampal system means remembering what happened the last time the person stood at the top of a tall building (his fear system was activated, his heart was pounding, he was terrified). When all those things come together in working memory—the stimuli that are present now, combined with the conscious memory of what happened the last time those stimuli were present and with the fact that the body is now very aroused (the unconscious memory is being reactivated)—that may be what feeling "afraid" is.

Emotion and Human Brain Evolution

Given that the amygdala has a much greater ability to influence the cortex than vice versa, what we're looking at here is the age-old debate between reason and passion. When you're aroused emotionally, whether by fear or by sexual attraction, for instance, your emotions dominate your thoughts. Philosophers and theologians going back to Plato and the ancient Greeks have talked about this seemingly fundamental schism, noting that the body fills us with passions and desires and fears, and all sorts of fancies and foolishness. For Plato, a true philosopher was someone who could gain control of his emotions by the use of reason, something Plato felt was a lifelong process. Socrates said, "Know thyself," by which he meant that we had to understand our emotions and be able to control them.

There have been other writers and philosophers through the ages who believed that in order to be truly human—as opposed to being in thrall to our animal nature—we have to learn to control our passions through reason. Descartes, of course, said, "I think, therefore I am." He didn't say, "I feel, therefore I am." Theodore Dreiser wrote: "Our civilization is still in the middle stage, scarcely beast in that it is no longer guided by instinct, scarcely human in that it is not yet wholly guided by reason." All of these thinkers aimed to put reason in charge and to minimize, if not eliminate, emotions. A contemporary example of someone who has managed that, of course, is Mr. Spock, of the *Star Trek* television series. Spock is a Vulcan who has no emotions and is pure reason. Herman Melville's protagonist in *Moby-Dick* was of the opposite persuasion. "Ahab never thinks," Melville wrote, "he just feels, feels, feels."

But perhaps there's an alternative to domination by either cognition or emotion. Maybe what we see in this imbalance between the amygdala's abundant input to the cortex and the relatively sparse cortical inputs to the amygdala is evolution in action. Even though thoughts can readily trigger emotions by activating the amygdala, we're not very effective at willfully turning off our emotions by somehow deactivating the amygdala. If we look at these connections across different species,

however, it is clear that the cortical connections to the amygdala are far greater in primates than they are in other animals. Perhaps as our brains evolve and these cortical pathways continue to increase, a balance might be struck. We obviously can't determine what's happening in this regard, because we can't watch our own brain evolving. But at least we can hope that this is the way it will go. A more harmonious integration of reason and passion in the brain would allow future generations of humans to know their true feelings and to use them more effectively in daily life.

7

Of Learning, Memory, and Genetic Switches

Eric Kandel

Who would we be without our skills and our memories? Our very identities are bound up in the historical sense we have of ourselves and the stories we tell about ourselves, in being able to say, "Last week I had dinner with my mother," or "For fifteen years I was captain of a nuclear submarine." We also define ourselves by the fact that we play tennis, say, but not the piano; that we can sing all the words to all the songs from our high school era but draw a blank on the names of the original performers; that we've always driven a stick shift but have never learned to dance. How do we learn and remember all these different things? What causes us to forget?

In this chapter, Dr. Eric Kandel, University Professor at Columbia University and Senior Investigator at the Howard Hughes Medical Institute, is one of the world's leading investigators in the study of learning and memory. He explains how the brain goes about creating memory of all kinds, from knowing how to ride a bike to knowing the capitals of all fifty states. The "switch" that transforms an event or fact or the mastery of riding on two wheels into something we can remember weeks or years later involves turning on the genes in brain cells that lead to the production of new proteins. As scientists discover more about how to manipulate this switch, Kandel predicts, we will have our hands on a powerful method for combating age-related memory loss. We will also gain a new understanding of how the brain creates—and continually re-creates—our memories and our selves.

IN THE PAST few years, scientists working in many different subdisciplines of biology, from evolution and genetics to immunology, biochemistry, and cellular neurobiology, have made discoveries that have revolutioned our understanding of mental processes and therefore our understanding of ourselves. As a result, when intellectual historians look back on the second half of the twentieth century, they will probably acknowledge that some of the most interesting insights of modern culture, and the most profound insights into the mind, will have come not from philosophy, literature, or the arts, nor even from psychoanalysis or psychology—disciplines that are traditionally concerned with culture and the mind—but from biology.

This increase in the explanatory power of biology comes from two developments. First, molecular biology has succeeded in unifying the biological sciences. For example, major advances in our understanding of the gene have enabled us to see how its double helical structure determines heredity even as its regulation (whether a given gene is turned on or off) determines cellular development and function. As biologists identify specific genes and the proteins they encode, their insights have given us a marvelous sense of the fundamental unity of biology, revealing an unanticipated similarity among proteins encountered in different species and in different cellular contexts. As a result, we now have a general blueprint for cell function that allows us to study various molecular processes in laboratory animals with confidence that our findings have a bearing on the human species as well.

A second, and equally profound, unification is that occurring between systems: neurobiology—the science concerned with the workings of the brain—and cognitive psychology—the science concerned with the working of mental processes. This unification offers new avenues for investigating a variety of mental functions such as perception, action, language, learning, and memory.

The two independent unifications at the extremes of the biological sciences—at the molecular level and at the level of mental processes—raise a question: To what degree can these strands themselves be united? Can molecular biology enlighten the study of mental processes, as it has enlightened all other areas of biology? Can we anticipate an even broader unification ranging from molecules to minds?

I would like in this brief paper to outline the possibility of a *molecular biology of cognition.* Using as examples several different forms of memory and learning in genetically modified snails, flies, and mice, I will try to illustrate that at least one component of mental processes—the switch from short-term memory to long-term memory, or how we convert a transient thought into an enduring memory—can now be studied by combining the tools and conceptual approaches of cognitive psychology with those of modern molecular biology.

Where in the Brain Is Memory Stored?

The problem of memory storage is commonly divided into two components. The first component is the systems problem:

Where in the brain is memory stored? Which systems in the brain are recruited for memory storage? The other component is the molecular problem: How is memory stored? What are the mechanisms whereby storage occurs?

The first question—where in the brain is memory stored?—has a long history in neuroscience, going back at least two hundred years, because it is part of a larger question—namely, to what degree are any mental processes localized to specific regions of the brain? Can aspects of language or thought or learning be said to take place in given regions of the brain?

The first person to address this question systematically was Franz Joseph Gall, a German physician and neuroanatomist who practiced and lectured in Vienna early in the nineteenth century. Gall made two major conceptual contributions to neuroscience. First, he did away with dualism, the idea propounded in the seventeenth century by the French philosopher René Descartes that the physical brain is a separate entity from the mind. Based upon his anatomical study of the brain, Gall argued that all mental processes emerge from the brain. "There is no soul," Gall argued, a pronouncement that later got him into difficulty with the Austrian authorities. Second, Gall was the first person to suggest that there was localization of mental functions to the brain. He was struck by the fact that the cerebral cortex did not have a homogeneous appearance; instead, its corrugations and wrinkles varied from one area to another. These physiological variations suggested to him that different parts of the brain might be concerned with different functions. Gall asserted that the brain does not act as a unitary organ but is divided into at least thirty-five organs (others were added later), each corresponding to a specific mental faculty or characteristic, including not only such things as memory or language but also musical talent, cautiousness, generosity, secretiveness, and romantic love (amativeness).

Moreover, Gall wasn't interested in simply describing the brain; he had a theory of how it worked. He thought that each mental property increased in size as a result of use, much as the size of a muscle is increased by exercise. For example, the part of the brain that is responsible for frugality would increase if a person saved a lot of money; similarly, the area concerned with clandestine activity would grow in someone who was very secretive. As each center for a given brain function grew, it would cause the overlying skull to protrude, creating a pattern of bumps and ridges on the skull that indicated which regions of the brain were most developed. By correlating the personality of individuals with the bumps on their skulls, Gall sought to develop an anatomical basis for describing character traits, a theory later called phrenology.

Gall was not an experimentalist, however. In particular, he did not think one could learn much from human patients with brain lesions or from experimental animals. In the 1820s, however, his ideas were subjected to experimental analysis by the French experimental neurologist, Pierre Flourens, who attempted to isolate the contributions that different parts of the nervous system made to behavior. In experiments with pigeons, he found that lesions of the cerebellum, at the base of the brain, destroy the animal's muscular coordination and its sense of equilibrium; and that damage to the medulla oblongata, at the back of the brain, results in death.[1] In the cerebral cortex, however, the region of brain which was the focus of Gall's attention, successive slicing through the hemispheres seemed to inflict damage on all of the higher mental functions equally, with the amount of damage to perception, intellect, and will varying with the extent rather than the location of the lesion.[2]

From these results, Flourens concluded that while sensory and motor functions are located in specific subcortical regions of the brain, higher mental functions such as perception, volition, and intellect are spread throughout the cerebral cortex and are not localized to specific regions of the cortex, the outer layer of the brain. In fact, Flourens suggested, the cerebral cortex functions equipotentially—that is, all areas participate equally in mediating all mental functions. Only when large regions of the brain are damaged do mental functions begin to suffer, and then the greater the damage, the more mental functions are interfered with, regardless of the specific region that has been damaged. This view, later called the aggregate field view of the brain, was rapidly accepted and dominated thinking in the first half of the nineteenth century.

In the second half of the nineteenth century, the situation changed dramatically, as people began to study the nature of language, the highest, and most characteristically human, cognitive function. It turned out that language function proved to be extraordinarily localizable.

The Structure of the Brain

To understand how language works, we need to look briefly at the brain's anatomy. In humans and other mammals, the higher mental processes are located in the brain's outer shell, the cerebral cortex, which is commonly divided into four regions, or lobes. The frontal lobe we now know to be concerned with motor coordination and with the planning of strategies and goals. The parietal lobe is concerned with speech perception and body sensations. The occipital lobe is concerned with vision. And the temporal lobe is concerned with hearing, smell, and, as you will later see, with aspects of memory storage. [Figure 13]

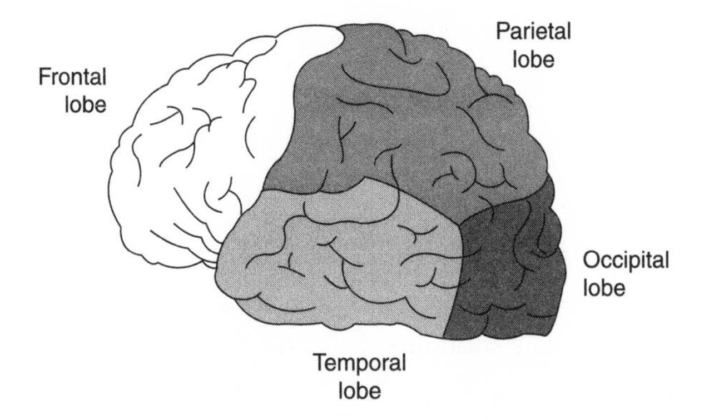

Figure 13 Most higher mental processes take place in the cerebral cortex, which has four lobes. The frontal lobe is active in planning, strategies, and goals as well as being involved with motor coordination. The parietal lobe is a major region for speech, perception, and body sensations. The occipital lobe is involved with vision, and the temporal lobe with hearing, smell, and aspects of memory storage. Adapted by Leigh Coriale Design and Illustration from *Brain, Mind, and Behavior,* 2nd edition by Bloom and Lazerson. © 1988 by Educational Broadcasting Corporation. Used with permission of W.H. Freeman and Co.

Most of what we know about the localization of language comes from the study of aphasia, a category of language disorders that result from medical problems. Aphasia is usually found in patients who have suffered a stroke, the obstruction or rupture of a blood vessel supplying a portion of the cerebral hemisphere. A number of important discoveries in the study of aphasia occurred in rapid succession in the last half of the nineteenth century. Taken together, these advances form one of the most exciting chapters in the study of human behavior, because they offered the first insight into the biological basis of a complex mental function.

That first insight came in 1861, from the work of Paul Broca, a neurologist working in France. Broca had a patient, called

Leborgne, who had an aphasia of a particular kind. He could understand language perfectly well, but he could not speak. He had no conventional motor problems with his tongue, mouth, or vocal cords. Indeed, Leborgne could whistle perfectly well, and he could hum a tune—he just couldn't sing the words. He could not speak grammatically or in full sentences, although he could utter isolated words. Moreover, the problem wasn't simply related to the articulation of speech; he couldn't write a letter, either. So Leborgne could not express language in any form.

When Leborgne died and an autopsy was performed, Broca examined his brain and found a lesion in the frontal lobe. He then became interested in this problem, and he followed several other patients who had similar symptoms. In each case, examination of the patient's brain after death revealed a lesion in the posterior region of the frontal lobe, and invariably it was on the left side. This area is now called Broca's area, and the disease is called Broca's aphasia. [Figure 14]

This discovery led Broca to announce, in 1864, one of the most famous principles of brain function: "*Nous parlons avec l'hémisphère gauche!*" ("We speak with the left hemisphere!") Only lesions of the left hemisphere cause this defect; Broca could show that people who had lesions in the opposite hemisphere, in the homologous region, didn't have this defect.

Broca's work showed that one could learn an enormous amount about which parts of the cortex were involved with cognitive functions such as language or perception by carefully studying patients who have cognitive deficits. His results stimulated a search for the cortical sites of other specific behavioral functions, a search that was soon rewarded. In 1870, two German scientists, the physiologist Gustav Fritsch and the psychiatrist Eduard Hitzig, galvanized the scientific community with their discovery that electrical stimulation of certain regions of a

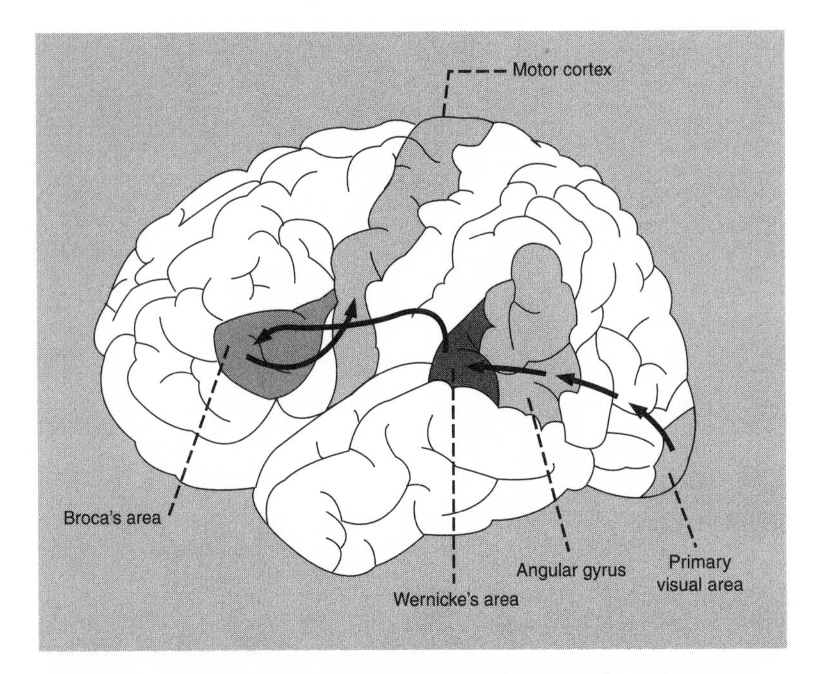

Figure 14 As scientists have learned from studying people with various forms of language loss, or aphasia, different language functions reside in different parts of the left hemisphere. In reading, the information comes into the brain through the visual cortex and is projected to Wernicke's area, where it is recognized as language and associated with meaning. Wernicke's area transforms the language from a sensory representation into a motor representation in Broca's area, which gives rise to the motor program that can lead to spoken or written language. Adapted by Leigh Coriale Design and Illustration from *Images of Mind,* by Posner and Raichle. © 1997 by Scientific American Library. Used with permission of W.H. Freeman and Co.

dog's brain produced characteristic movements of the limbs. They found that individual movements are controlled by small, quite discrete regions of the motor cortex. Moreover, the movements of limbs on one side of the body were controlled by the opposite side of the brain. Thus, in human beings, the right hand, commonly used for writing and other skilled movements, is controlled by the left hemisphere, which also, as we have seen, controls speech. In most people, therefore, the left hemisphere is regarded as being dominant.

Things became even more exciting a few years later, in 1874, when a German neurologist by the name of Carl Wernicke found another patient with aphasia, but an aphasia that was the opposite of the one Broca had found. Broca had found a patient who could understand but could not speak. Wernicke found a patient who could speak but could not understand; he couldn't even understand his own speech. This man could say words perfectly well, but they made no sense,[3] and he could not understand what was said to him. When he died and came to autopsy, he also had a lesion in the left side of the brain, but in a different site from Broca's area. This lesion was in the posterior part of the temporal lobe, where it joins the parietal and occipital lobes, a region now called Wernicke's area.

But Wernicke's contribution didn't stop with describing an isolated syndrome. In addition to making this discovery, he developed a model of the language system in humans. As the basis of his model, Wernicke proposed that only the most primary mental functions, those concerned with simple perceptual and motor activities, are localized to single areas of the cortex. More complex intellectual functions, he said, result from interconnections among several functional sites.

Wernicke thus advanced the first evidence for the idea of *distributed processing,* which governs the modern view of mental functions. That is, mental functions are indeed localized in the brain, but important functions are usually not localized to a single region. Instead, important functions are typically broken down into various components, and different components of the functions are represented in different areas of the brain. This profound insight—the complexity of localization in the brain—is the guiding principle of modern neural science.

Wernicke next proposed a model of language that though highly simplified is still operative today. Neurologists still use it at the bedside to make 90 percent of their diagnoses. The model works as follows: When you read or hear language, the information comes into your brain through the primary sensory areas of the cortex specialized for visual (reading) or auditory (hearing) information. These areas then project to Wernicke's area, where the information is recognized as language and associated with meaning. Without that association, the ability to comprehend language is lost. Wernicke's area transforms the comprehended language from a sensory representation into some sort of motor representation in Broca's area that can lead to spoken or written language. Thus Broca's area forms the motor program for the grammar that gives rise to the articulation of language as speech.

Based upon this model, Wernicke then predicted a new type of aphasia. He predicted that an aphasia could occur without damage to either Broca's or Wernicke's area by simply interrupting the pathway between the perceptive area and the motor area. The resulting disconnection syndrome, now called conduction aphasia, is characterized by an incorrect use of words. These patients understand words that are heard or seen, and they can talk, but they cannot speak correctly. They omit parts of words, or substitute incorrect sounds. There's no relationship between their understanding and what they say.

Inspired in part by Wernicke, a new school of cortical localization arose in Germany at the beginning of the twentieth century, led by the anatomist Korbinian Brodmann. This school sought to distinguish different functional areas of the cerebral cortex based on the structure of cells and the characteristic arrangement of these cells into layers. Brodmann distinguished fifty-two areas in the human cerebral cortex that could be associated with some sort of identifiable function, primarily sensory and motor areas.

Given this progress, and continued interest in finding brain regions that deal with higher cognitive functions, it was only a matter of time before the interests of neurobiologists converged on memory. Where is it stored? Is there a specific location in the brain for memory storage? And if so, are all memories stored in the same place?

The person in the United States who is generally identified with early efforts to localize memory is Karl Lashley, a professor of psychology at Harvard and perhaps the dominant figure in American neuropsychology in the first half of the twentieth century. Lashley explored the surface of the cerebral cortex in the rat and systematically removed different cortical areas. In doing so, he failed repeatedly to identify any particular region of the brain that was special to, or necessary for, the storage of memory. So Lashley—harking back to Flourens—formulated the law of mass action, according to which the extent of the memory defect is correlated with the size of the cortical area removed, not with its specific location.

However, Lashley focused his efforts only on the cerebral cortex, the outer layer of the brain; he never explored structures that lie deep in the cortex. As it turns out, subsequent work has shown that many forms of memory require one or another subcortical region. Furthermore, the exercise that Lashley's rats performed in his experiments involved learning to navigate a maze, a task that recruits many motor and sensory functions. When a cortical lesion deprived an animal of one set of cues (such as tactile cues), it could still learn the maze with vision or olfaction, so the experiment proved nothing about memory per se.

The first clear suggestion that aspects of memory might be stored in specific regions of the brain came from the work of the neurosurgeon Wilder Penfield at the Montreal Neurological Institute. Penfield was a student of the British neurophysiologist Charles Sherrington, who had used electrical stimulation to map how movement was represented in the motor region of the cerebral cortex of anesthetized monkeys. In 1938, in the course of his pioneering neurosurgical treatment for the relief of focal epilepsy, in which the neurons' electrical storm is confined to a circumscribed area of the brain, Penfield began to use electrical stimulation to map motor and other behavioral functions in the cortex of his fully conscious patients. The brain has no pain receptors, so once the scalp was anesthetized patients could remain conscious and describe what they were experiencing. The technique is still used today, because surgeons must identify specific brain sites that are important for language in the individual patient so that they can avoid these sites when removing epileptic tissue.

As Penfield performed these surgeries, and in the process explored most of the cortical surface in more than a thousand patients, he found that stimulation in certain areas of the brain—the temporal lobes—would occasionally cause people to have a sort of déjà vu experience, a faint memory of something that seemed to have happened before. He called them experiential responses. Patients would vaguely remember hearing a song that was familiar to them, or having a hazy vision of somebody they might know, or of being in a familiar context.

This work led Penfield to begin to think that memory could be localized to the medial temporal lobe, but his view was questioned by many. First, all of the patients he examined had abnormal brains due to focal epilepsy, and in 40 percent of the cases the mental experience evoked by stimulation was identical to the mental experience that ordinarily accompanied the patient's seizures, including elements of fantasy, as well as improbable or impossible situations. In fact, the experiences were actually more like dreams than memories.

However, other researchers, stimulated by Penfield's findings, eventually showed that the temporal lobes, along with the hippocampus, which lies deep within them, were critically important in human memory. The most dramatic case was that reported in 1957 by William Scoville, a surgeon, and Brenda Milner, a psychologist and a student of Penfield's. They reported the extraordinary study of a patient known as H.M.

When he was nine years old, H.M. was hit by a bicycle and suffered a brain injury that left him with epilepsy. The condition worsened during the next eighteen years, until H.M. was too incapacitated to continue his work in an assembly plant. His seizures couldn't be controlled with medication. At the age of twenty-seven, as a last resort, H.M. underwent an experimental surgical procedure in which Scoville removed parts of both sides of the medial temporal lobe and the underlying hippocampus. The surgery succeeded in relieving the seizures, but it left H.M. with a devastating memory loss. From the time of his operation in 1953 until today, H.M. has not been able to convert a new short-term memory into long-term memory.

Milner, who discovered and described this memory defect, has observed H.M. for more than forty years. She found his memory loss to be surprisingly selective. That is, the surgery seems to have interfered with certain components of memory but not with others. By noting which aspects of memory function H.M. lost and which he retained, Milner was able to delineate four features that characterize the role of the temporal lobes and the hippocampus in memory storage.

First, H.M. had a reasonably good long-term memory for events that had occurred long before the operation. He could remember vividly many events of his childhood and his later experiences at work. He could still speak English coherently and fluently. His overall intelligence was essentially unchanged. All of these findings indicated that the temporal lobes and the hippocampus are not the ultimate storage site for long-term memories of previously acquired knowledge or experience.

Second, he had a perfectly good short-term memory. He could immediately repeat a new telephone number as accurately as anyone else. He could remember a person's name when he was first told it. He could carry on a normal conversation, provided it didn't last too long or involve too many topics. So the temporal lobes and the hippocampus are not required for short-term memory, either.

Third, what H.M. lacked, however—and lacked to an extraordinary degree—was the ability to put new information into long-term storage, so he appeared to forget events almost as soon as they happened. Less than an hour after eating, he could not remember anything he had eaten, or even that the had had a meal. He would read the same magazine over and over without knowing that he had read it. As years passed, he could not recognize himself in a photograph because he had no memory of his changed appearance. If he was distracted, even briefly, after being given a number to repeat, he forgot the number completely. He did not recognize new people, even when he met them repeatedly. Other patients with lesions of the hippocampus show similar learning deficits, as do experimental animals with such lesions. Milner therefore showed that lesions of the temporal lobes and the hippocampus led to a dissociation of short-term memory from long-term memory—a finding that provided the first evidence that there is more than one memory system.

The last characteristic Milner discovered was that even this defect for converting information from short-term memory into long-term memory was not absolute. In fact, there were certain types of learning that H.M. could carry out and remember perfectly well. Specifically, he was still able to learn new motor skills that are not dependent on awareness or cognitive processes. For example, he could trace the outline of a star while looking at a reflection in a mirror; that is, he was not allowed to look directly at what he was doing but had to watch his hand in the mirror. Even though he would not remember from one day to the next that he had ever done this task before, his performance improved daily, as would that of a normal subject.

What these and subsequent studies revealed is that memory has at least two major forms. One form is concerned with *knowing how,* or knowledge of motor skills; the other is concerned with *knowing that,* or knowledge of facts and events. "Knowing how"—also called nondeclarative or implicit memory—does not require conscious awareness. "Knowing that"—also called declarative or explicit memory—requires a conscious focusing of attention. Declarative memory involves memory for facts, ideas, and events that can be brought to conscious recollection. Nondeclarative memory is a change in behavior that occurs as the result of experience, where memory is expressed through performance and without conscious recollection of past episodes. Moreover, the two forms are localized in different systems of the brain. Declarative memory involves the medial temporal-lobe system, including the hippocampus. Nondeclarative memory involves specific sensory and motor pathways in the brain as well as the cerebellum, the amygdala, the basal ganglia, and some other structures.

The two major forms of memory, declarative and nondeclarative, can also be subdivided. Declarative memory can be divided into two components, called episodic and semantic. Semantic memory concerns general knowledge of facts and vocabulary and is thought to involve the hippocampus. Episodic memory is concerned with the recollection of the context of specific events—where they occurred, for instance, or the time of year—and is thought to be represented in part in the frontal cortex. Nondeclarative memory is subdivided into forms of learning that include classical conditioning (the process Pavlov used to train his dogs to salivate at the sound of a bell), as well as ordinary habits and learned skills such as touch-typing or playing tennis.

The Mechanisms of Memory Storage

Now that we know that memories are stored at specific sites in the brain, it becomes interesting to turn to the second component of memory: the storage mechanisms. Given that there are two major systems of memory in the brain, and that they use completely different logic and completely different brain systems, it becomes of further interest to ask to what degree the two systems have storage mechanisms in common. Can molecular biology reveal similarities?

One clue to shared mechanisms has come from studying the stages in memory storage. Memory for both declarative and nondeclarative forms of learning is commonly divided into at least two stages: a short-term memory that is very unstable and lasts only minutes and a long-term memory that is stable and self-maintained and can last for days, weeks, or even years. In both declarative and nondeclarative memory, repetition is commonly required to convert short-term memory to long-term memory. A recent important finding is that in both forms of memory, the transition from short- to long-term memory requires a spurt of new protein synthesis. Now, proteins are important in everything we do—with every thought, every dream, every action, we're using a supply of proteins that are there all the time. When we throw the switch to translate something from short- to long-term memory, however, we need a spurt of *new* proteins. Once the memory has been acquired—once it's been in a stable store for more than an hour—memory can maintain itself by using the proteins that are already available.

This dependence of new protein synthesis was first demonstrated by Louis Flexner of the University of Pennsylvania, Bernard Agranoff of the University of Michigan, and Samuel Barondes and Larry Squire of Albert Einstein Institute. Agranoff demonstrated the mechanism using goldfish. He trained goldfish in a particular task until he knew exactly how many training trials were needed for them to perform 100 percent accurately twenty-four hours later. Let's say that 100 trials were necessary. Agranoff trained scores of goldfish on 100 trials, then put them into different groups and injected them, at various times after learning, with drugs that inhibit protein synthesis. For example, he took ten fish and injected them immediately after learning. Another ten were injected after half an hour, ten more after an hour, and so on. The results showed that the inhibitors have no effect on learning and memory when injected two hours after training. But if they are injected immediately after training they block memory storage dramatically.

The experiments demonstrated that the inhibitors of protein synthesis have no effect on learning per se and no effect on short-term memory, but they selectively block long-term memory. Further, the requirement for protein synthesis has a specific time window called the consolidation phase. That is, long-term memory is most sensitive to disruption during this consolidation phase—during and immediately after training. If exposure to the protein-synthesis inhibitor is delayed by as little as one hour after training, animals experience no significant deficit in long-term memory. Moreover, other studies have shown that the requirement for protein synthesis during the early phase of memory consolidation is evident not only in vertebrates but also in the fruit fly, *Drosophila,* and the marine snail, *Aplysia.* This generality suggests that some of the key proteins that make up the switch might also be common across species and used for both declarative and nondeclarative forms of memory storage. If this is so, identifying the relevant proteins in one species might yield insights into the memory systems of other species.

Nondeclarative Memory in *Aplysia*

We therefore began our inquiry into the molecular nature of memory by studying the protein-synthesis-dependent step in the marine snail called *Aplysia.* This creature is a good subject because its nervous system contains only approximately twenty thousand neurons, compared with the billions of nerve cells in the mammalian brain. Moreover, *Aplysia's* nerve cells are huge, the largest cells in the animal kingdom; some are even visible to the naked eye. These cells are also distributed in ten different regions of the brain called ganglia. A ganglion contains about two thousand cells and controls not one but a family of behaviors. As a result, the number of cells committed to a single behavioral act is very small, about 100 cells or fewer. Our studies showed that *Aplysia* exhibits several forms of nondeclarative memory. We have reason to believe that the switch from short- to long-term memory used by this animal is very general and might work the same way as in other animals, and in the human brain.

Aplysia has a simple reflex response in which it withdraws its gill when its fleshy spout, the siphon, is stimulated: If you touch the siphon lightly, the animal withdraws its gill, much the way a human would snatch a hand back from a hot object. [Figure 15] This reflex can be modified by different kinds of learning processes, including the form of nondeclarative learning called sensitization. In this process, the animal learns to recognize an offensive stimulus and to strengthen its defensive reflex response to stimuli that were previously neutral. So, for example, when *Aplysia* is presented with a noxious stimulus to the tail, such as an electric shock, it learns to speed up its reflex responses in preparation for withdrawal and escape. Once it has been sensitized in this way, *Aplysia* will show a much more powerful withdrawal of the gill to subsequent innocuous stimulation of the siphon than it would have done previously.

As is generally the case, the memory of this sensitizing stimulus varies with repetition. A single noxious stimulus to the tail gives rise to a short-term memory that lasts only minutes and does not require the synthesis of new protein. But five tail stimuli result in a long-term memory that lasts several days and does require protein synthesis. Subsequent training gives rise to an even more persistent memory.

So, depending upon the number of repetitions, sensitization can give rise to both short- and long-term memory. Stimulating the tail activates a serolonergic modulatory system within the animal which triggers a chemical signaling system within the sensory neuron called the cyclic AMP (cAMP) system. With a single stimulus, or pulse of serotonin, the cAMP signaling system works primarily in the watery cytoplasm of the sensory cell, where it activates a special protein (the cAMP-dependent protein kinase) that regulates the function of other proteins. The result is an increased release of the neurotransmitter glutamate, which strengthens communication between the sensory neurons and the motor neurons and the interneurons for a pe-

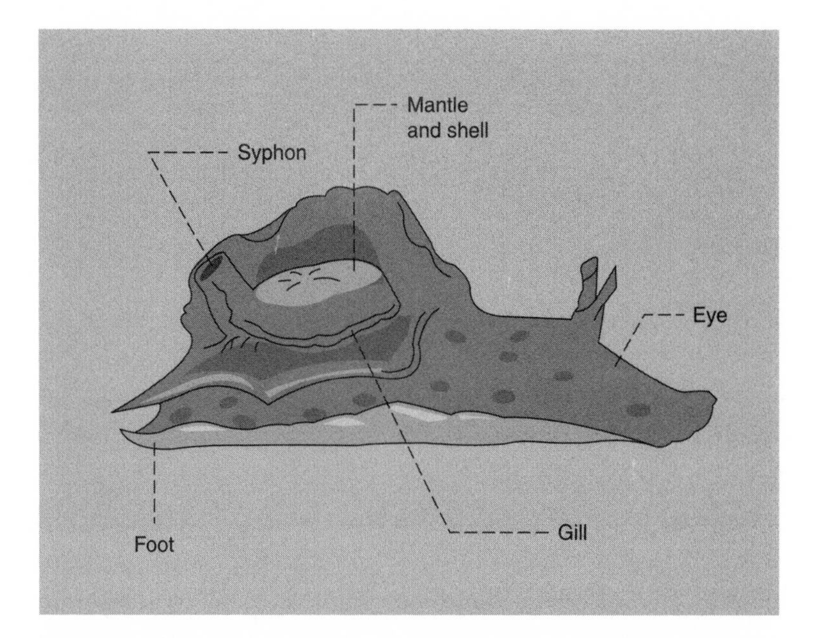

Figure 15 The sea snail *Aplysia* has a simple reflex response that makes it a good subject for studying learning and memory. When its fleshy spout, the siphon, is touched lightly, the animal withdraws its gill. If its tail is then given a more obnoxious stimulus, such as an electric shock, *Aplysia* speeds up its reflex responses in a classic fight-or-flight reaction, and it becomes much more reactive to subsequent light stimulation of the siphon. Adapted by Leigh Coriale Design and Illustration from *Molecular Cell Biology,* 3rd edition by Lodish, Baltimore, Berk, Zipursky, Matsudaira, and Darnell. © 1995 by Scientific American Books. Used with permission of W.H. Freeman and Co.

riod of minutes. That's the cellular representation of short-term memory.

With repeated training, a component of this signaling system moves the special protein (the cAMP-dependent protein kinase) into the nucleus of the cell, where it acts on genes to synthesize proteins. This is the genetic switch; in order to remember something in the long term, genes must be switched on.

Not surprisingly, given the complexity of memory, the action of this genetic switch for long-term memory is complex. It has "excitatory elements," gene activators (called CREB-1), which bind to the regulatory region of a gene to turn the gene on to produce a protein, which in turn creates the growth of synaptic connections. It also has "inhibitory elements," or gene repressors (CREB-2), which sit in the regulatory regions of the genes and are designed to prevent expression of the gene. So at the very earliest stages of switching on the gene, there is surprisingly powerful control. Not only must the CREB-1 activator be activated but the CREB-2 repressor, which is preventing CREB-1 from acting, must be eliminated.

Why is this important? We've all experienced dramatic evidence of this process from time to time. Sometimes we seem to be able to remember things—to transfer them to long-term memory—with the greatest of ease. And at other times we can't remember the name of someone we met at a dinner the night before. One reason is that the repressor might be interfering with long-term memory storage. What about the corollary? If one could remove the repressor, would there be instantaneous long-term memory? Dusan Bartsch, who discovered this repressor in my lab, tested the idea by developing a specific antibody that would get rid of the CREB-2 repressor in *Aplysia.* Bartsch found that getting rid of the repressor did indeed produce instantaneous long-term synapic facilitation. With just one trial, one pulse of serotonin, plus the antibody to CREB-2, *Aplysia* demonstrated an enhanced reflex strength that lasted twenty-four hours, which was indistinguishable from the results of five training trials.

A similar enhancement of memory occurs in human beings as well. When we're in a situation in which our emotions are heightened—waiting for a decision or a letter to arrive, for instance, or getting ready to walk down the aisle to get married—anything we experience under those circumstances, such as what the person sitting next to us is wearing, we are likely to remember for a longer period of time than we remember anything that happens during the course of an ordinary day. One probable reason is that the regulatory system involved in removing the repressor has been activated; we are thus prepared to recall things effectively. This enhancement could happen just through random gene mutation; random variations of repressor proteins undoubtedly exist in the normal population. In other words, some people who seem to have an extraordinary memory may just have defective repressors. This is not necessarily a condition to be envied. Such people complain that they can't forget anything—in effect, entire pages of useless telephone numbers are forever inscribed in memory, and they see everything through a constant screen of prior images, or as a kind of afterimage.

What purpose does all this genetic machinery of activators and repressors for long-term memory serve? The short answer is: It allows neurons to grow new synaptic connections while they stabilize long-term memory. In recent years, we have learned that what maintains a memory is a growth of new synaptic connections; any kind of learning literally alters the functional connections in our brains. [Figure 16] When *Aplysia* was trained with the sensitization task, for example, both its sensory neurons and its motor neurons grew new connections. The same thing probably happens in the human brain. The parts of the cortex that are involved in motor and perceptual skills are also thought to be modified anatomically in this way.

A cartoonish way of depicting the way experience changes the mammalian brain is to "illustrate" the body surface—the skin—on a map of the cortex that shows which parts of the cortex correspond to touch sensation. If we drew the map of the skin as a kind

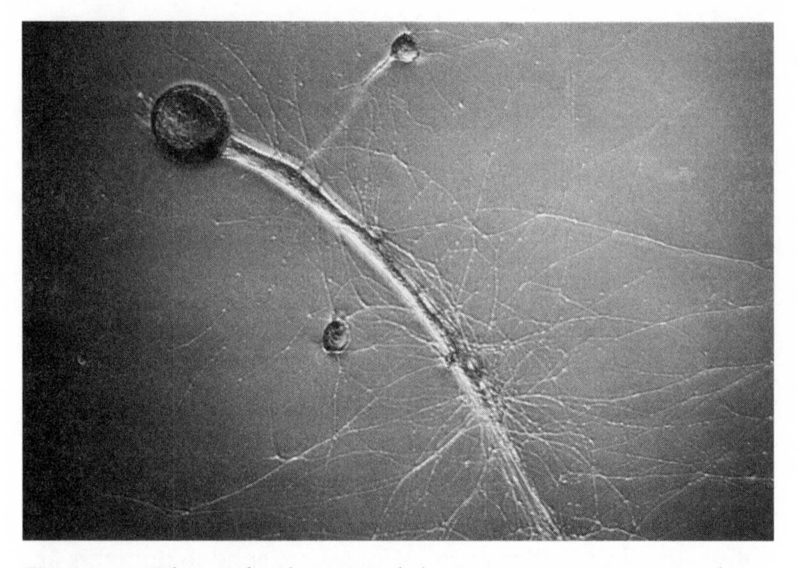

Figure 16 When *Aplysia* learns new behavior, its sensory neurons and motor neurons both grow new synaptic connections. Similar growth probably occurs in the human brain in response to any kind of learning or training. Image courtesy of Doctors Kelsey Martin and Eric Kandel, Howard Hughes Medical Institute, NY.

of miniature body, or homunculus ("little man"), as such a depiction is called, and attempted to show each part of the body as larger or smaller depending on how much of the cortex is devoted to it, the result would be a very distorted little man. For example, the hands and the face are extremely sensitive tactile organs, and would have a much larger representation in the cortex (and, hence, on the homunculus) than the back or chest. [Figure 17]

In a classic study of how learning affects the brain, Mike Merzenich of the University of California, San Francisco, has looked at the representation of the hand area on the cerebral cortex in monkeys. When he examined different monkeys at random, he found that their hand representations varied. To discover if this was simply genetic variation or whether and to

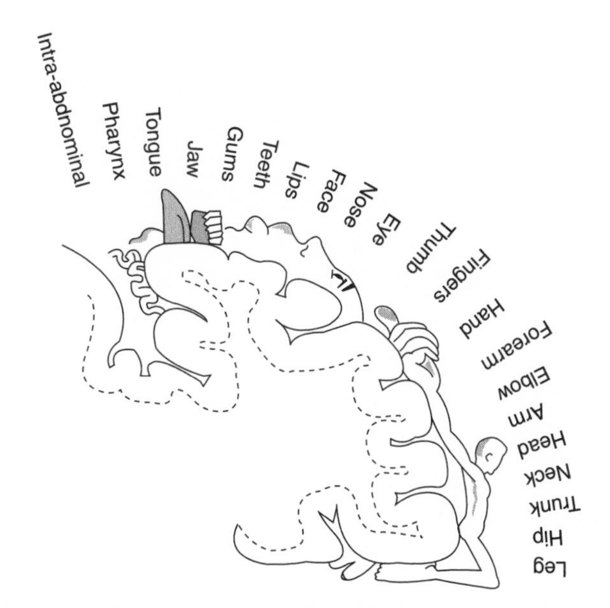

Figure 17 This distorted homunculus, or "little man," represents the touch-sensitive surface of the body as a map on the cortex, with larger and smaller areas corresponding to the amount of the cortex devoted to it. For example, the hands are extremely sensitive and have a much larger representation than, for instance, the back. Adapted by Leigh Coriale Design and Illustration from *The Brain: A Neuroscience Primer,* 2nd edition by Thompson. © 1993 by W.H. Freeman and Co. Used with permission.

what degree experience played a role, he did an experiment: He gave one monkey the equivalent of piano lessons. The monkey was taught to press a bar for food using only the fingertips of three fingers. The training went on for several weeks. When Merzenich examined the monkey's cortex later, the area representing those three fingers had expanded at the expense of other areas of the brain.

This is a significant finding. All of us—even identical twins that share the same genes—are brought up in uniquely different social environments. We interact with different people, have different experiences in the world. As we learn from each of these experiences, the genetic switches in our brain cells are turned on or off, producing structural changes that make each of our brains unique.

What does this mean from a philosophical point of view, or from a clinical point of view? First, it means that the distinction that psychiatry has traditionally made between so-called functional mental illness, meaning the disorder does not arise from abnormal physiology, and organic mental illnesses, meaning that the illness has a biological cause, is nonsense. Everything is organic.

This is not to say that these illnesses can't respond to nonphysiological treatments such as psychotherapy. Speaking to a counselor or a friend may help, in fact. But the key point to remember is that even therapy helps by retraining—and thus altering—the brain.

Insofar as any stable change produced in our behavior involves structural changes in the brain that are produced by throwing genetic switches, then all illnesses have a genetic component. Some are inheritable and some are not. For example, a disease like schizophrenia or major depression—or diabetes, for that matter—has a strong heritable genetic component. If your parents have had a major mental illness such as schizophrenia or depression, there is a greater likelihood that you or your children will also have that illness. This is because the several genes that predispose an individual to various mental illnesses occur in every cell of the body, including the sperm and the egg, and are, therefore, passed from generation to generation.

This principle of genetic inheritance does not apply to brain changes that result from the activation of genes in the process of learning. That is, a frightening experience such as falling off a horse may send a signal that ultimately will turn on certain genes in the brain cells that were off initially; those genes will now trigger production of a protein that may make you feel anxious. This anxiety is determined by genes in the sense that genes were switched on and off. But it will not be passed on to your children because it is not a structural change in your DNA. Moreover, it occurs in only a few select cells of your brain. It does not occur in your spleen, your heart, or your gonads; it will not be present in the egg or the sperm.

Modern imaging techniques are beginning to make it possible for us to observe how brain changes are precipitated by various kinds of environmental stimuli and learning. On the one hand, PET scans, for example, reveal that the brains of people who are skilled in a particular task consume less energy—are more efficient—than those of people who are just learning it; the experienced brains have generated new synaptic connections, speeding the task performance along. On the other side of the coin, functional magnetic resonance imaging has shown decreases in volume in the hippocampus of people who have suffered severe stress or trauma; imaging also shows, however, that this neuronal damage and hippocampal shrinkage is reversible, provided the stress doesn't go on too long. More dramatically, very young children who have undergone surgery to remove fully half of their brains in order to control potentially lethal epileptic seizures have recovered substantial amounts of normal brain function, because neurons in the remaining half of the brain went into overdrive to compensate.

Given the brain's plasticity—this ability to rewire itself in response to environmental stimuli and any kind of learning—and the continued advances in our understanding of how this rewiring occurs at the molecular level, modern imaging techniques are opening up exciting possibilities for both the treatment and the diagnosis of various illnesses and disorders. For example, we may be able to discover what underlies age-related memory loss, the inability to translate information from short-term memory into long-term memory. If, as seems likely, the problem lies in a defective genetic switch, preventing the necessary activator/repressor interaction and production of new proteins, research could conceivably come up with a pill to overcome the problem; we could then test the results with brain imaging.

Finally, the recent advances along all these fronts in neuroscience and molecular biology should make one thing clear. Those who may have thought that the intrusion, as it were, of biology into psychology, into higher mental processes, would somehow dehumanize the human psyche—or at least banish its wonder and reduce it to an uninteresting machine—are wrong. If anything, just the opposite is the case: The greater our understanding of the brain, the more amazing it becomes. Just because we know that the heart is a pump, for example, doesn't mean we think it's no longer a wonderful organ; understanding its structure and biology hasn't diminished its magic. And so it is with the brain. Now that we begin to see some of the brain's molecular underpinnings, we can only be more awestruck by the fact that the action of all those molecules somehow results in our being able to read a book or write one, fall in love, fall down laughing, learn to tie our shoelaces, and remember our fourth-grade teacher's name many decades later. A lifetime of learning and memory, encoded in the constantly renewed connections between and among our brain cells, makes us who we are.

8

Order from Chaos

J. Allan Hobson

Freud called dreams the "royal road" to the unconscious, believing that the study and interpretation of dreams—a kind of top-down approach—would yield insights into a person's deepest memories and feelings. Dr. J. Allan Hobson, a professor of psychiatry at Harvard Medical School and the director of the Laboratory of Neurophysiology at Harvard, approaches our bizarre nighttime narratives from the bottom up, looking instead at the neurochemical characteristics of waking, sleeping, and dreaming. His research suggests that dreams, and especially their hallucinatory character, are largely the product of two physiological events—the spontaneous activity of neurons in the brain stem, which fire without benefit of external stimulation,[1] combined with a drop in the neurotransmitters that pump through the more rational parts of our brains during waking hours. Lacking the chemical juice of rationality, our meaning-making cortex nonetheless tries its best to turn the assorted brain-stem signals into a sensible story, with results that can leave us amused or bemused—or wake us up gasping in terror.

As WITH so many aspects of our understanding of the human brain and mind, generally accepted notions of sleep and dreaming have evolved over time. The ancient Greeks, for example, believed that sleep was controlled by a winged god, Hypnos, who was the son of Nyx, the god of the night, and the brother of Thanatos, the god of death. Hypnos lived in a cave with his three sons: Morpheus, who brought dreams to humans; Icelus, who brought dreams to animals; and Phantasus, who brought dreams to inanimate objects. The Greek view, then, was that an outside agent caused both sleep and dreams. Later, the Christians held on to the idea of an outside agent but converted the winged gods of classical antiquity into angels, who became the emissaries of God. In the medieval worldview, God was the source of dreams, which were thought to convey important messages to humans about how to behave.

By now, we've shifted the emphasis away from external agents to look inside the brains of humans and our fellow mammals for clues to what dreaming is and how it works. We've been making excellent progress. In fact, what we've learned recently has unalterably changed the view of dreaming that has prevailed since Sigmund Freud published *The Interpretation of Dreams* in 1900. Freud wanted to establish a dream theory that was, as he put it, "perspicacious and free from fault," by basing it on brain research. The level of research necessary to create this kind of theory simply did not exist in Freud's time, however, and is only now developing.

In looking at the dreaming brain, the emphasis here will be on neuroscience rather than on the psychological aspects of dreaming that were Freud's focus. That is, we won't be delving into dream interpretation, even though the meaning of dreams has been an irresistible topic for people of every culture and era throughout the ages. Indeed, one of the problems with efforts to approach the psychology of dreams in a scientific manner is that we can't actually do the science on dreams themselves because we have no means of spying directly on the movie being played in the mind of the dreamer. Instead, we must rely on the dreamer's subsequent reports. And dream reports, by virtue of the fact that the dreamer must put whatever images and sensations he or she can recall into words, have already been interpreted, are already literature. As the eighteenth-century poet John Dryden put it in "The Cock and the Fox":

Dreams are but interludes
Which fancy makes.
When monarch reason sleeps
This mimic wakes,
Compounds a medley of disjointed things
A mob of cobblers, a court of kings.

In what follows here, we'll be looking at the underlying physiology of the brain in trying to find answers to the questions How is it that reason sleeps? What is the "mimic" that wakes? and What is the purpose or use of the "medley of disjointed things"? For example, dreams have an uncanny ability to persuade us that we're awake, a delusion that persists in the face of extraordinary, even impossible, events. In dreams, we may find ourselves soaring over mountaintops like Superman, or having deep conversations with famous, long-dead personages. So we'll be examining what might give rise to the visual experiences, the strong emotions, and the hallucinatory quality of dreams. In other words, what changes in the brain could account for the presence of these remarkable features and the reciprocal loss of rationality?

The experience of intense emotion is one of the things that distinguish dreaming from most waking life. Moreover, as systematic surveys of normal people's dreams show, most dreams are unpleasant; in fact, the dominant emotion is anxiety or fear. What this suggests is that the parts of the brain that deal with fear, the pathways that include the hypothalamus and the amygdala, are almost certainly selectively activated during the stage of sleep that is associated with dreaming. This makes sense from an evolutionary point of view, if we consider how vulnerable an animal is when it is sleeping. A certain amount of anxious vigilance would be an important survival tool, provided it was kept just below the threshold of consciousness so as not to be incapacitating. In humans, anxiety is the number-one dream emotion. Number two is elation, but third, and way ahead of everything that follows it, is anger. So, of the top three, two unpleasant emotions combined, anxiety and anger, far outweigh the pleasant emotion of elation. Interestingly, other emotions that are markedly underrepresented in dreaming are shame, guilt, and sadness. This distribution of emotional content in dreams may be telling us something about which emotions are innate and which are conditioned by society.

In addition to intense emotion and loss of rationality, dreaming is characterized by fairly severe memory loss. That is, even though we dream for at least an hour and a half to two hours every night, we're lucky to recall even a few minutes' worth of our dreams. So when "monarch reason" sleeps, we lose not only our capacity to think and analyze but also the capacity to remember. Upon waking, a dream may seem disjointed and disorienting not only because the dream narrative shifted nonsensically but also because there are gaps where we simply can't recall what happened next.

Dreaming is not the only state in which we suffer disorientation and loss of memory, however. Amnesiacs, for example, or people with Alzheimer's disease, don't know where they are or what's going on around them. Dreaming might, in fact, be a kind of model for some forms of mental illness, such as schizophrenia, in which disorders of memory are suspected of being at the root of the cognitive disturbance.[2] In that sense, to use Freud's phrase, dreaming may indeed be "the royal road" to understanding the mind, both normal and abnormal. Dreaming itself is not pathological, but it is a kind of psychotic condition—a normal psychotic condition, as it were—that affects all of us every night of our lives. If we can begin to get a fix on how dreaming works, we will have a blueprint for beginning to understand how the symptoms that are normally confined to sleep may arise when we are awake, enabling us to reach into the mental hospitals and into the minds of afflicted fellow humans.

The Physiology of Dreaming

Since the 1950s, we've known that most dreams occur during periods of sleep characterized by rapid eye movement, or REM. Indeed, early dream researchers concluded that dreams occurred *only* during REM sleep because most people who were awakened during this stage of sleep reported being in the middle of a dream. However, dreams can occur in non-REM (NREM) sleep as well, although these dreams tend to be briefer and less extraordinary in content. The new brain-based theory of dreaming looks at the parts of the brain that are active during waking periods, NREM sleep, and REM sleep. For example, when we're awake, and "monarch reason" reigns, the cerebral cortex, the most recently evolved part of the brain, is extremely active. It is in this part of the brain that our higher cognitive functions such as language, memory, and planning occur. Parts of the cortex are also involved in such functions as motor output and sensory perception. During NREM sleep, the cortex is the part of the brain that is largely inactivated, which accounts for the poverty of mental activity during that stage of sleep. In REM, many parts of the cortex are reactivated, but not the frontal regions responsible for working memory (where we hold information temporarily) and higher-level functions such as planning and decision making. No wonder we become disoriented and unable to control our thoughts when dreaming.

Somehow, then, certain parts of the brain are, in effect, turning themselves on and off during sleep and dreaming. Our studies suggest that the mechanism by which the brain turns itself on and off is chemical. That is, the chemical state of the brain during dreaming sleep is distinct from its chemical state during waking periods.

Whether we're sleeping or awake, the brain automatically maintains certain basic physical activities that are essential to life, such as breathing, heartbeat, and blood pressure. All of these are primarily the responsibility of the brain stem, a three-inch-long structure at the base of the brain, where it meets the spinal cord. Deep in the center of the brain stem is a region called the pons. Higher cognitive functions in the cortex are turned on by a sustained activation process that arises in the pons. Some repetitive motor functions, like walking and running, also begin with signals sent from the pons down the spinal cord.

When we're awake—and have reason, logic, and memory—some cells in the pons are continuously secreting certain neurotransmitters called amines, such as norepinephrine and serotonin. These substances are drastically reduced during REM sleep, while another chemical system—the cholinergic system, generated by another cluster of cells in the pons—kicks into high gear, putting out quantities of the neurotransmitter acetylcholine in amounts equal to those found in the waking brain. Acetylcholine is a neurochemical that is found throughout the brain and the body. It is the major neurotransmitter at the junction between nerves and muscle;[3] among other things, it slows the heart, constricts the pupils of the eyes, and causes salivation and blushing.[4] During REM sleep, acetylcholine excites the neurons that inhibit muscle action, move the eyes, and activate the upper brain.

Much of what we know about the physiology and neurochemistry of sleep and dreaming has been learned from studying not only human subjects but also our fellow mammal the cat, because cats and humans have analogous sleep patterns. As illustrated on the following page [Figure 18], for example, electrical traces of both people and cats during REM sleep show similar levels of activity for eye movement, muscle tone, and brain waves. The low-voltage, rapid brain-wave pattern during REM is actually similar to the pattern an observer would see if the person or the cat were awake. But the nearly flat line representing muscle tone is unique to REM sleep. We know that some parts of the motor system are activated during REM sleep because the eyes move. However, the rest of the motor system is inhibited; in human and cat, the muscles of the body are flaccid and effectively paralyzed.

As anyone who has been around infants or very old people knows, individual sleep patterns can vary quite a bit, depending on an individual's age. Level of fitness and types of daily activity are also factors in determining how soundly or how many hours a night one sleeps. Generally speaking, however, during the course of the night we go through regular cycles of alternating

REM SLEEP

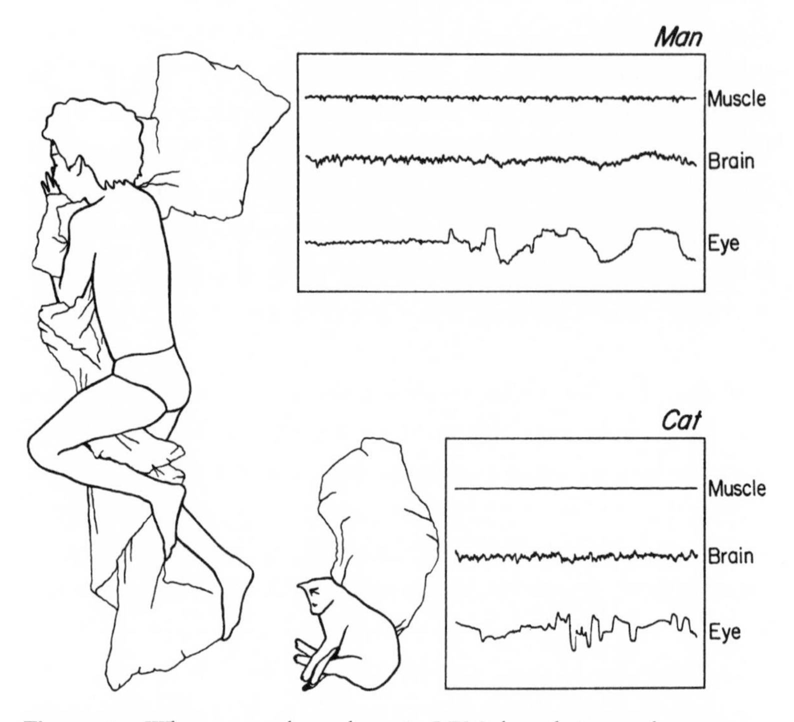

Figure 18 When cats and people are in REM sleep their muscles are essentially paralyzed but their rapid brain wave activity looks much as it does when they are awake. Illustration courtesy of Dr. Allan Hobson, Harvard Medical School.

REM sleep as well as the four-stage process of brain deactivation that characterizes NREM sleep. During NREM sleep we are deeply unconscious, especially early in the night, when we linger longest in the deepest stage of NREM, Stage IV. During this stage, the brain's electrical activity is at its most disorganized. Not only are awakenings difficult but the person awakened tends to be very confused.

The Cycles of Sleep

Once we fall asleep and have been asleep for about 60 or 80 minutes, the level of activity in the brain starts to rise as we drift back up from Stage IV through Stage III on to Stage II and Stage I. Instead of waking up at that point, however, the rapid eye movements begin. After some period of REM, usually lasting about ten minutes for the first pass,[5] the brain starts to drift back down through the four stages, and then cycles back up. This process repeats itself every 90 to 100 minutes. (The cat's sleep cycle is also regular, but it lasts about 30 minutes instead of 90.) As the night progresses, the REM periods get longer and the NREM periods become shorter and shallower. Instead of descending all the way down to Stage IV NREM sleep, we tend to dip only to Stage III and then climb back up to Stage II and an REM period. By morning, the distinctions between REM and NREM are further diminished as we alternate between Stage II and Stage I REM. [Figure 19].

The activated brain periods—the REM periods—occupy between 20 and 30 percent of every singe night of sleep. So if we sleep for eight hours, then roughly 25 percent, or two hours, of that time will be spent in a state of brain activation associated with hallucinatory dreams and emotional intensification. In an average life span of seventy years, the amount of time spent dreaming comes to something on the order of fifty thousand hours, or nearly six years. The fact that we spend so much time in REM sleep would seem to indicate that evolution has placed a high priority on putting the brain into this state; something profoundly important to our biology must be going on for nature to devote so much time to this.

Now, as we all know, when we're dreaming, we see things; the visual imagery is detailed and intense. It is, in fact,

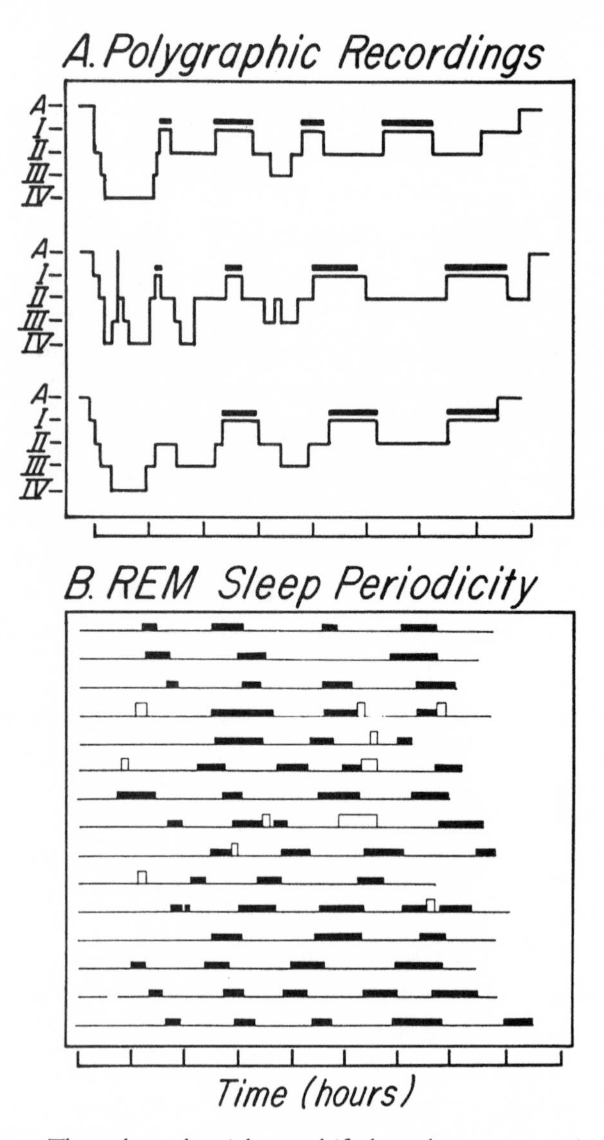

Figure 19 Throughout the night, we drift through 90- to 100-minute cycles in which REM sleep (the solid black bars) is interspersed with four stages of non-REM, or NREM, sleep. We spend 20 to 30 percent of each night in REM sleep, dreaming an average of two hours per night. Illustration courtesy of Dr. Allan Hobson, Harvard Medical School.

hallucinatory. However, the visual cortex, the part of the cortex that normally gets input from the eyes, does not receive external stimuli from the eyes, because the eyes are closed. Moreover, we know that even people who have lost their sight due to accident or illness later in life continue to dream in the visual domain. Therefore the visual cortex in the dreaming brain must be stimulated from some source other than the eyes.

Research with cats has helped to determine exactly which parts of the brain are active and when. What we've learned is that the stimulation that occurs during REM sleep comes from the brain stem, and specifically from the pons. The messages these cells send to the visual cortex are responsible for our hallucinations. Indeed, two specific clusters of cells in the pons seem to act as "on" and "off" switches for REM sleep. The cluster that produces acetylcholine triggers REM; the clusters that produce norepinephrine and serotonin bring REM to a halt. Acetylcholine cells are called cholinergic in contrast to the aminergic norepinephrine- and serotonin-containing cells.

When we're awake, the brain cells that produce norepinephrine and serotonin are active, and we can line up our thoughts, think logically, and process external data. We know what time it is, and where we are. Our feelings are under some degree of rational control. But when the cells that produce these neurochemicals turn off during dream states, rational judgment disappears. In effect, we believe any foolish thing that our neural signals send, and we assume that it's real.

When we measure the electrical activity of one neuron in the pons of the cat brain during REM—a neuron that may be producing acetylcholine—we get recordings like the one shown on the next page. [Figure 20] Even though the cat's outward appearance is still that of tranquil sleep, an electrical storm is raging inside its brain, and we record enormous spike and wave

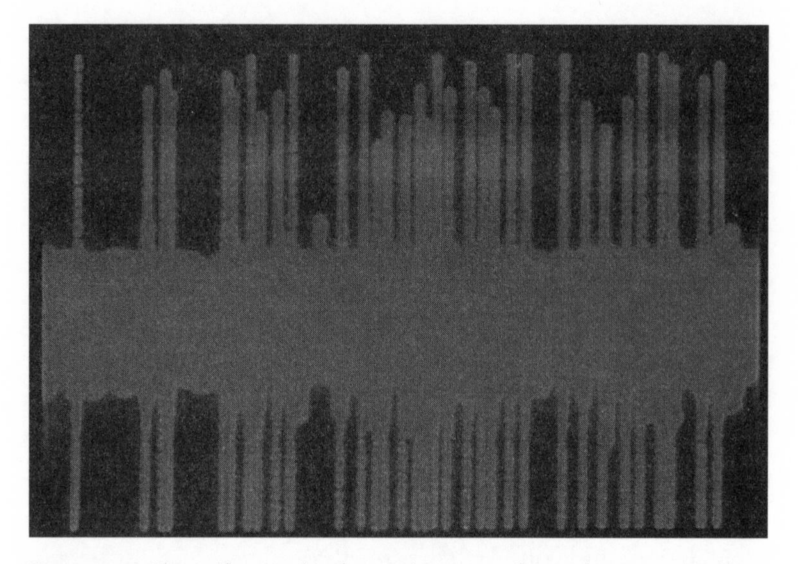

Figure 20 Shown here is the electrical activity of a single neuron in the pons of a cat during REM sleep. A veritable electrical storm is occurring in the brain of an animal that is lying still. Illustration courtesy of Dr. Allan Hobson, Harvard Medical School.

complexes that greatly resemble the brain waves that are typical of an epileptic seizure. The cells in the pons are firing spontaneously and wildly, broadcasting signals called action potentials all through the brain; by exciting occularmotor cells in the brainstem, they cause the eyes to move. Pontine cells also mediate muscle inhibition, which results in a loss of muscle tone and virtual paralysis, giving rise to the classic nightmarish feeling of wanting to run and of being frozen in place. (Although one might have thought sleepwalking is associated with REM sleep and dreaming, it is actually a disorder that is associated with non-REM sleep. In effect, it is a disorder of arousal: The individual's cortex is deep in the throes of Stage IV, but somehow enough activation begins down in the lower brain regions to support walking.)

The signals from the pons also travel to the emotional brain—especially to the amygdala—generating the anxiety that typifies our fight-or-flight response. Finally, the barrage of signals that began in the pons radiates up to the cerebral cortex, where memories and input from the senses normally come together. But because the aminergic cells have been turned off, "monarch reason" is hampered in its usual cognitive duties—the lack of neurotransmitters such as norepinephrine and serotonin upsets the normal assembly and association process that goes on during waking states.

One could say, then, that dreams are the product of our cortex's efforts to do the best it can under very difficult operating circumstances. Indeed, this is a credit to the extraordinary creative capability of the brain. As electrical signals travel through the brain, triggering memory fragments and spurious sensory input, the cortex pulls them together into stories and visual images. This, of course, could be the explanation for the inspirational nature of dreams, especially those that are related to artistic and scientific discovery.

We can get a sense of how other dream images and sensations might be produced by looking at the operation of a nerve cell in another part of the brain, the cerebellum. The cerebellum, or "little brain," located at the base of the brain, is the part of the brain that is concerned with coordinating movement and the memory of motor procedures. Making up nearly 10 percent of the brain's weight and containing nearly half its neurons,[6] the cerebellum is the brain structure that allows us to learn to ride a bike at the age of eight and remember how to do it at the age of fifty-eight. Cells in the cerebellum fire continuously throughout periods of waking and sleeping and are very active during REM sleep, even though actual changes in posture and most movement (except the eyes and some facial twitching) are inhibited.

Instead, we experience fictive movement and imaginary posture changes in our dreams, the constant animation that typifies most dreams. Not surprisingly, given what we know about the cerebellum's role during waking states, neurobiologists tend to think all this REM activity in the cerebellum must have an influence on the development and maintenance of the motor system and motor memories.

Dream Reports

It is one thing to measure physiological changes in the brain during REM sleep and quite another to investigate the subjective experience of dreaming, a pursuit that relies on oral or written dream reports. Most people can't resist interpreting their dreams, ascribing meanings and associations to the night's images and events. Occasionally, however, some rare individuals can give detailed reports that are of scientific use, reports that are free of interpretation. One particular informant, an entomologist who worked in the Department of Agriculture, began a dream journal in the summer of 1939. For 100 successive nights, he recorded, in copperplate handwriting, 254 dream reports. These reports constitute an important database for our work, because they were written in a dispassionate manner, with descriptive detail that is free of interpretation. In addition to writing his matter-of-fact verbal descriptions, the scientist also illustrated his dream journal—a rare find indeed.

Intending the journal only for his own use, he wrote, and drew, in a completely naive and unguarded fashion. In the following illustration [Figure 21], for example, he depicts a dream of flying on a magic carpet. He describes his sensations as he flies through the dream space: "I am floating in the air . . . about

Figure 21 From the dream journal of a scientist comes a depiction of a flying carpet—a pleasant sensation of weightlessness that may be prompted by random stimulation of the brain's position-sense neurons. Illustration courtesy of Dr. Allan Hobson, Harvard Medical School.

fifteen or twenty feet from the floor . . . There is a moderate wind blowing and this moves me slowly forward." Internal stimulation of the brain's position-sense neurons, located in the brain stem, may be the source of this experience of weightlessness.

In another image, the scientist gives us his rendition of a nightmare. He is hanging by his fingertips from railway tracks on a high trestle. [Figure 22] Baggage carts up on the tracks are about to run over his fingers—a terrifying situation. This kind of terror is familiar in dreams. Anxiety is automatically stimulated when spontaneous signals from the pons reach the amygdala; the cortex, upon receiving signals from the amygdala, then generates the imagery that is designed to fit the emotion.

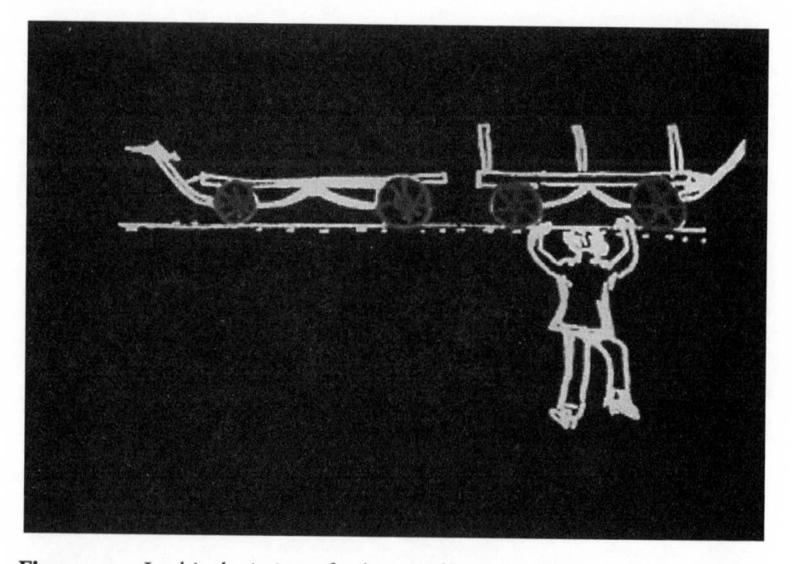

Figure 22 In this depiction of a dream of hanging from a railroad trestle, the dreamed movement is terrifying rather than pleasant, perhaps because the fear systems in the amygdala have been stimulated as well. Illustration courtesy of Dr. Allan Hobson, Harvard Medical School.

There's another kind of nightmare, called a hypnagogic hallucination, in which we awaken from a dream and a dream character is still in the room. What seems to be going on is that the cortex, though partially awake, is still being bombarded by stimuli from the brain stem, so we're really in two worlds at once. Fortunately, this is usually a short-lived event. The human brain and mind are normally quite capable of making a sharp distinction between these two worlds.

However, the fantastic quality of dreams, and especially the intense emotions of nightmares, seems ripe for psychological interpretation. Freud, of course, was convinced that dreams (if we could only decode them) were the key to revealing unconscious motives and inarticulate feelings. He believed that primitive appetites of the sort we share with animals—for sex, food, and dominance—were kept in check by a higher part of the mind (the ego). Indeed, the ego not only kept these urges in check but essentially kept them hidden from us, beneath our conscious awareness. However, if these urges were not acted upon (or "resolved," as Freud termed the process), they would find another outlet—hence dreams. Freud thought that dreams were the mind's way of expressing sexual and violent urges by disguising them in symbolic imagery; they're disguised because if they were allowed to invade consciousness they would disrupt sleep. The dreamer has difficulty remembering the dream, Freud maintained, because these urges were forbidden material.[7]

Modern scientists tend to resist Freudian dream interpretation. For us, dreaming is simply the subjective awareness of the product of the brain's spontaneous activation during sleep. Dreaming has its peculiar character—its discontinuities and incongruities, its disorientation and its hallucinatory quality, its intense emotionality—all of which we can attribute to the distinctive activation pattern and the distinctive chemical state of the brain during REM.

This is not to say that dreams have no meaning, however, or that they are entirely nonsensical, although they do contain a great measure of nonsense. Indeed, as chaotic as they often are, dreams are full of meaning. They are the brain's effort to make sense of a nonsensical situation, and as such a dream is a bit like a Rorschach test. The spontaneous firing of neurons creates cognitive strain, and the cortex—every individual's particular cortex—has to fill in the blanks. Where neuroscientists and Freudian psychologists may disagree is over the idea that images in dreams are, necessarily, symbols that disguise some hidden or buried meaning, and that dream symbols have meanings that are universal. We neuroscientists contend that dreams are not, as Freud would have it, an effort to disguise hidden sexual impulses—indeed, they seldom do a good job of disguising those impulses at all. Instead, dreams reveal interesting aspects of the psyche.

The Purpose of Dreams

So if dreams are not necessarily symbolically disguised messages from the repressed unconscious, what are they? The seats of memory and reason in the cortex are inactivated during dream states, as is the neurochemistry needed to make the cortex function properly. Why should the brain do this?

One answer is that REM serves to aid brain development. We know that REM sleep is highly overrepresented in early life—infants, for example, spend about twelve hours a day in REM sleep. In adults, the activation of the brain during dreaming sleep causes the feeling of movement and the experience of emotion; in ways that we don't yet understand, all of this is probably in the service of fostering memory and development. For example, an intriguing aspect of REM sleep is that the part of the brain called the hippocampus, which has been associated with memory, is as active during REM sleep as it is when awake animals explore new environments. Indeed, some animal studies suggest that neuronal activity during REM sleep reinforces activities related to survival that the animal carried out when it was awake, such as exploratory behavior in a rat or predatory behavior in a cat.[8] (It may be that dreams are so powerfully disorienting because the other thing the hippocampus seems to do is help animals orient in a real space. But, of course, in REM sleep the hippocampus is turned on but there's no external space to orient to; the only space that you're orienting to is memory space.)

There's some good experimental evidence for the theory that REM sleep fosters long-term memory consolidation. In one study, for instance, subjects are exposed to a visual learning task that requires considerable attention and can be learned with difficulty in a couple of hours. The subjects are then allowed to sleep as usual, and when they are retested the next morning their performance is as good as, and sometimes even better than, it was when they stopped taking the test the night before. But if they're deprived of REM during that sleep and retested the next morning, their performance is no better than it would have been if they were taking the test for the first time.

Thus even though dreams themselves cannot be remembered, they seem to enhance remembering. This suggests that the cognitive content of dreams may be a kind of incidental by-product of brain activity during REM sleep, with relatively low survival value, since we can't remember our dreams when we wake up. By contrast, the brain process that supports dreaming is essential to higher cognitive functioning.

We know, for example, that REM deprivation in humans is deleterious. The working hypothesis when REM sleep was first discovered was that it was necessary for maintaining mental health. Hence those early REM-deprivation studies were seriously misguided, because researchers expected the REM-deprived subjects to go crazy, and some of them did. After five days of REM deprivation, many individuals become psychotic and paranoid; they hear voices, and their ability to function declines drastically. This is also true, however, of people who are deprived of non-REM sleep, so the specificity of those early experiments is somewhat questionable.

Recently, however, the processes that support REM sleep have been shown to be essential to life. Experiments by Allan Rechtschaffen of the University of Chicago show that rats deprived of REM sleep long enough die, and they die from a disorder of metabolic control. That is, they cannot maintain their body weight, and they can't control their body temperature. This is interesting, because REM sleep is the only time in the life of the mammal when temperature control is abandoned; that is, we do not thermoregulate in REM sleep. Ambient conditions must therefore be optimal—we must have shelter and a warm bed—in order for us to have REM sleep. If the ambient thermal conditions are adverse, sleep is fitful and REM sleep is prevented.

The animal experiments have yielded another finding. One of the ways in which we can deprive animals of REM sleep—in addition to waking them up, as is done with humans—is to administer certain chemicals. If we overstimulate the aminergic system, as with amphetamines or other stimulants, we shut down the REM system. When this happens, the animals incur a REM debt; the longer they are prevented from having REM sleep, the more intense is the debt payback.

In humans, this can happen with alcohol and certain drugs, which, in sufficient quantities, suppress REM sleep and create a REM sleep debt. Thus when the addict can no longer get the drug and goes into withdrawal, then the REM debt is paid—and it's paid off with interest. The intensity of the REM process that occurs under this debt-payback circumstance is psychotogenic—it causes the hallucinations and tremors of delirium tremens and probably contributes to the psychosis of withdrawal from such drugs as amphetamines and cocaine.

The chemistry of sleep and dreams is a burgeoning field. Now that we have a specific set of hypotheses about how the system works—the functions of the various neurotransmitters in mediating REM and NREM sleep—we can actually turn REM on and off in the laboratory, and we're learning much that will increase our understanding of dreams and the dreaming brain. As we've seen, REM sleep is essential, putting our brains through paces that seem to enhance the consolidation of memory and learning. And although the subjective experience of the dreams that occur during REM may not necessarily have survival value, the emotional quality of our dreams—whatever we can recall of them—offers each of us insights into our own concerns.

If anything, our dreams offer further proof, if any is needed, that the three-pound organ housed inside each of our skulls has developed its own unique neuronal connections, giving rise to our individual personalities and our personal inventory of talents, memories, fears, and joys. Triggered by our brain's own spontaneous signals, we produce narratives that are often meaningless and sometimes meaningful, but that meaning is ours alone. Jonathan Swift, who is best known as the author of *Gulliver's Travels,* anticipated all of this in a poem called "On Dreams," which he wrote on 1727:

> Those dreams that on the silent night intrude,
> And with false flitting shapes our minds delude,
> Jove never sends us downward from the skies,
> Nor can they from infernal mansions rise,
> But all are mere productions of the brain,
> And fools consult interpreters in vain.

NOTES

Chapter 1

1. Report of the National Institute of Mental Health's genetics work group, September 1997.

2. Hyman, quoted in "Mental Illness Is No Myth," by Tom Siegfried and Sue Goetinck, *Dallas Morning News,* 1996.

3. "Would a Child of Mine Have Schizophrenia?" by Clea Simon, *Washington Post,* March 9, 1997.

4. Op.cit. Report of the National Institute of Mental Health's genetics work group.

5. "Next Generation of Psychiatric Drugs . . . ," *New York Times,* November 19, 1996.

6. "Fertile Minds from Birth," *Time,* February 3, 1997.

7. Ibid. Also, "Gene Involved in Brain Development," Rockefeller University press release, April 17, 1996.

8. Op.cit. "Fertile Minds from Birth."

9. Op.cit. Report of the National Institute of Mental Health's genetics work group.

Chapter 3

1. Edward Thomas, *The Icknield Way* (London: Constable, 1913), pp. 280–283.

2. Quoted in Allan Seager, *The Glass House: The Life of Theodore Roethke* (Ann Arbor: University of Michigan Press, 1991), p. 101.

3. Ian Hamilton, *Robert Lowell: A Biography* (New York: Random House, 1982), p. 157.

4. *Byron's Letters and Journals,* 12 vols., ed. Leslie A. Marchand (London: John Murray, 1973–1982), (letter from Byron to Earl of Clare, February 6, 1807), vol. 1, p. 106 and (letter from Byron to Edward Noel Long, April 16, 1807), vol. 1, p. 114. Hereafter cited as *BLJ.*

5. *BLJ,* (letter from Byron to Francis Hodgson, October 12, 1811), vol. 2, pp. 111–112.

6. Quoted in Malcolm Elwin, *Lord Byron's Wife* (New York: Harcourt, Brace & World, 1962), p. 256.

7. Leslie A. Marchand, *Byron: A Biography,* 3 vols. (New York: Alfred A. Knopf, 1957), p. 1112.

8. Vladimir Mayakovsky, "It's After One," lines 1–12, in Edward J. Brown, *Mayakovsky: A Poet in the Revolution* (New York: Paragon House, 1988), p. 356.

9. John Berryman, "384: The Marker Slants," lines 1–10, *The Dream Songs* (New York: Farrar, Straus and Giroux, 1969), p. 406.

10. Edgar Allan Poe, *The Letters of Edgar Allan Poe,* vol. 2 (letter to Annie L. Richmond, November 16, 1848), ed. John Wand Ostrom (Cambridge: Harvard University Press, 1948), pp. 401 and 403.

11. George Gordon, Lord Byron, *Byron's Letters and Journals,* vol. 5, *1816–1817* (letter to Thomas Moore, January 28, 1817), ed. Leslie A. Marchand (Cambridge: Belknap Press of Harvard Universiy Press, 1976), p. 165.

12. John Nichol, *Byron* (New York: Harper & Brothers, 1880), p. 12.

13. MacKinnon, D. F., K. R. Jamison, and J. R. DePaulo, "Genetics of Manic-Depressive Illness." *Annual Review of Neuroscience,* 20: 355–373, 1997.

Chapter 4

1. Sternberg, E., and P.W. Gold, "The Mind-Body Interaction in Disease," *Scientific American* special issue, "Mysteries of the Mind," 1997, p. 12.

2. Ibid., p. 10.

3. "Skin-Deep Stress," by Mike May, *American Scientist,* May–June 1996.

4. "Tracing Molecules That Make the Brain-Body Connection," by Elizabeth Pennisi, *Science,* vol. 275, no. 5302, February 14, 1997, pp. 930–31. The last paragraph quotes McEwen about neuroendocrine immunology being a holistic approach, a coming together of fragmented sciences: "We're putting the body back together again."

Chapter 5

1. *The Defending Army* (Time-Life Books, *Journey through the Mind and Body* series, n.d.), p. 102. Also mentioned in "Accentuate the Positive," by Emma Haughton, *The Independent,* December 3, 1996.

2. Op.cit., Pennisi, "The Brain-Body Connection."

3. This section (pp. 104–109) based on "Emotions and Disease" exhibition catalogue by Theodore M. Brown; Exhibition Directors Elizabeth Fee and Esther M. Sternberg; curators Anne Harrington and Theodore M. Brown. Copyright 1997, Friends of the National Library of Medicine.

4. *Random House Dictionary of the English Language,* 2nd ed. unabridged.

5. *Defending Army,* p. 102.

6. Op.cit., Pennisi, "The Brain-Body Connection."

7. *Defending Army,* p. 31.

8. Op.cit., Pennisi, "The Brain-Body Connection."

9. Ibid.

10. *Mind and Brain,* (*Journey through the Mind and Body* series), p. 33.

11. Op.cit., Pennisi "The Brain-Body Connection"; *Mind and Brain,* pp. 50–51 for definition of vagus nerve.

12. Sternberg, E., and Philip Gold, "The Mind-Body Interaction in Disease," *Scientific American* special issue, 1997, p. 14.

13. Ibid.

14. *Defending Army,* p. 105.

15. Ibid. pp. 105–106.

16. Op.cit., Sternberg and Gold.

Chapter 7

1. "Flourens, Marie-Jean-Pierre" Britannica Online. http://www.eb.com:180/cgi-bin/g?DocF=micro/212/86.html

2. Mind, Brain, and Adaptation, the Localization of Cerebral Function; http://www.tau.ac.il/~yosiba/adapti.html. 22 Aug. 1996.

3. *Mind and Brain* (*Journey through the Mind and Body* series), p. 18.

Chapter 8

1. Hobson, J. Allan, *Sleep* (New York: Scientific American Library, 1995), pp. 17–18.

2. "The Prefrontal Cortex and Schizophrenia," from the Society for Neuroscience, 1995.

3. Restak, Richard, *Receptors* (New York: Bantam Books, 1995), pp. 23–25.

4. Hobson, J. Allan, *The Chemistry of Conscious States* (Boston: Little, Brown and Company, 1994), p. 265.

5. "The Meaning of Dreams," Jonathan Winson, in *Scientific American* special issue, 1997, p. 59.

6. *Mind and Brain* (*Journey through the Mind and Body* series), p. 56.

7. *Secrets of the Inner Mind* (*Journey through the Mind and Body* series), pp. 11–12, and Hobson, *Sleep,* p. 146.

8. Winson, "The Meaning of Dreams," pp. 61, 63.

Index

acetylcholine, 31, 185, 189
ACTH, 85, 86, 100
action potentials, 130, 190
activators, gene, 172, 178
adaptation, 84, 86–87, 88
addiction, 4, 10, 22, 23–27, 198
Ader, Robert, 120
adolescent mood disorder, 59
adoption studies, 23
adrenal glands, 84–85, 87
adrenaline, 84–85, 89, 92, 96
adrenocorticotropic hormone, 85, 86, 100
aggregate field view, of brain, 156
aggression, 49
Agranoff, Bernard, 168
AIDS, 113
Akiskal, Hagop, 71
alcohol, 100, 107
alcoholism, 4, 22, 24–25, 198
Alexander, Franz, 106
allostasis, 86–87
allostatic load, 87–89, 93, 99, 100
alpha brain-wave pattern, 44
Alzheimer's disease, 98, 113, 183
amenorrhea, 88
amines, 184–85, 189, 191, 198
amino acids, excitatory, 96, 98
amnesia, 183
amphetamines, 198
amygdala, 115, 166
  dream activity role, 182, 191, 193
  emotional memory and, 94–95, 140–42, 146, 147
  emotions and, 125, 136, 148–49
  fear-response role, 132–35, 138, 139, 145
  personality development research, 34–35, 41–42, 45–46, 50
Andreasen, Nancy, 70–71
angels, 180
anger, 182
anorexia nervosa, 88
antibodies, 112
antigens, 91–92, 110
antisocial behavior, 49
anxiety, 182, 191, 193
anxiety disorders, 49, 126
aphasia, 157–58, 160, 161
Aplysia (snail), 169–74
approach behaviors, 45
aroused behavioral profile, 36
arthritis, 104, 116, 118–19
aspirin, 113
asthma, 82, 83
atherosclerosis, 88, 90, 91
atypical depression, 118
auditory sensory system, 131–32

autism, 18
autonomic nervous system, 84–85, 87, 94
axons, 12–13, 14, 130

Bangs, Lester, 124
Barondes, Samuel, 168
Bartsch, Dusan, 172
basal ganglia, 166
battle fatigue, 106
Berryman, John, 64
biphasic hormonal action, 83
bipolar illness. *See* manic-depressive illness
black bile (humor), 31, 105
blood (serum), 110, 111–12
blood (humor), 2, 31, 105
blood-brain barrier, 112
blood pressure
  fear and, 128, 129
  stress and, 82, 83, 88, 90
body temperature, 112, 198
body type, 47–48
bone marrow, 91, 110
Bowman Gray University, 90
brain, 2–4, 151–78
  addiction mechanism, 24–27
  asymmetry in function, 43–44
  dream activity and, 7, 180–99
  emotions and, 18, 124–49
  environmental effects on, 18–20, 175–78
  fear system and, 126–27, 129–35, 144
  fetal development of, 17, 18
  immune system and, 94, 110, 111–17
  learning effects on, 15, 19–20, 173–76, 177
  memory storage by, 151–52, 153–56, 162–69
  mental illness and, 9–12, 24, 73, 74, 77–78
  neuron structure and function, 12–18
  personality determinant research, 43–46
  plasticity of, 3, 7, 14–20, 94, 177
  stress effects on, 5, 82, 84–85, 94–99, 108–10, 115
  structure of, 156–67
  tracing connections in, 130
brain imaging, 11, 97, 108–9, 177
brain lesions, 129–32, 155, 158
brain maps, 129–30, 163
brain stem, 184, 189, 190, 193, 194
Breuer, Josef, 48
Broca, Paul, 43, 157–58, 160
Broca's aphasia, 158
Broca's area, 158–61
Brodmann, Korbinian, 161–62
Byron, Ada, 69
Byron, Lady, 60
Byron, Lord ("Mad Jack"), 69
Byron, Lord (poet), 59–61, 66, 67–69
Byron family, 67–69

cAMP system. *See* cyclic AMP system
cancer, 24, 99, 103, 121
capillaries, 112
cardiovascular disease, 64, 83, 88, 90–91
cardiovascular system, 87, 88, 89–91, 94
caregivers, 93, 119
cataracts, congenital, 19
catecholamines, 89
cells, 16–17 (*see also* immune cells; neurons)
central nervous system, 5–6, 12, 110, 111–15
cerebellum, 155, 166, 191–92
cerebral cortex, 17, 19, 173–75
  cognition and, 137, 154, 155–63
  dream activity and, 7, 179, 183–84, 191, 193–95
  emotions and, 131–35, 145, 148–49
  personality development research, 43–45, 50
  structure of, 156–57
cerebrum, 17
Chatterton, Thomas, 64

children's personality prediction research, 32–51
cholinergic system, 185, 189
chromosomes, 16
chronic fatigue syndrome, 118
classical conditioning, 120, 127–32, 139, 167
CNS. *See* central nervous system
cocaine, 25–26, 27, 198
cognition, 3, 5, 74, 137, 153–63
cognitive psychology, 153
Cohen, Nicholas, 120
colitis, 83
concentration, 31, 59
conditioned reflex, 127
conditioning, psychological, 120, 127–32, 139, 167
conduction aphasia, 161
conscious memory, 140–42, 145, 147
consciousness, 1, 3, 7, 143–47, 166
coping, 84, 94, 99–101
corticosteroids, 111
corticotropin releasing hormone, 85, 86, 110, 113–19, 121
cortisol
  hypothalamus and, 34, 85, 142
  immune system and, 113–16, 118
  stress and production of, 6, 34, 84–89, 92, 93, 96–98, 100
creativity, 5, 54–79
CREB-1 and CREB-2, 172
CRH. *See* corticotropin releasing hormone
cultural values, 47
Cushing's syndrome, 97–99
cyclic AMP system, 170–71
cyclothymia, 58, 71
cystic fibrosis, 22
cytokines, 88–89, 104, 111–13

Davidson, Richard, 44, 45
death, 75, 82–83 (*see also* suicide)
declarative memory, 166–69
déjà vu experience, 163
Delayed Type Hypersensitivity, 91–93
delirium tremens, 198
dementia, 59, 98, 113, 183
dendrites, 12, 13–14, 16, 96, 97
dentate gyrus, 95–96, 97
deoxyribonucleic acid. *See* DNA
depression, 10, 22, 45, 118, 176
  in creative people, 53–79
  immune system and, 93, 104, 118–19
  loss of concentration from, 59
  onset and episodic pattern of, 58
  stress and, 98, 99, 100
Descartes, René, 2, 148, 154
Dhabhar, Firdaus, 91–93
diabetes, 83, 89–90, 176
diet, 18, 89, 100
Dilantin, 97, 98
disease. *See* health and illness
displacement behavior, 94
distressed behavioral profile, 36
distributed processing, 160–61
diurnal rhythm, 100
DNA, 16, 17
dopamine, 14, 17, 25–26, 31
dreams, 6–7, 179–99
  dream reports, 181, 192–95
  physiological mechanisms, 183–86
  purpose of, 196–99
  sleep cycles and, 187–92
  symbolism of, 7, 195–96
Dreiser, Theodore, 148
drug abuse. *See* substance abuse
drugs
  gene activation/deactivation and, 20, 21
  immune system and, 113, 119
  as mental illness therapy, 3, 54, 75–77
  neuron damage and, 97, 99
  placebos, 107–8
  REM sleep and, 198
Dryden, John, 181
DTH. *See* Delayed Type Hypersensitivity

dualism, 2, 154
Dunbar, Helen Flanders, 106

Eastern Europe, 83, 90
economic costs, of stress, 83
ECT (electroconvulsive therapy), 54
ectomorph body type, 48
EEGs (electroencephalograms), 44
ego, 195
elation, 182
electrical brain stimulation, 163
electroconvulsive therapy, 54
electroencephalograms, 44
emotional memory, 6, 124, 138, 140–47
emotions, 3, 123–49
    brain function and, 18, 124–26, 147–49
    disease and, 5–6, 103–21
    dreams and, 181, 182, 191, 193–94, 199
    feelings vs., 125, 143–46
    unconscious, 133–37
    See also specific emotions
endocrine system, 104, 110
endomorph body type, 48
environment, 1–5, 10, 111
    genetic interrelationships with, 4–5, 22–24, 27–28
    as learning factor, 175–76
    neuronal change and, 14–15
    as personality determinant, 31, 32, 40–41, 47, 48–49
    stress and, 99–100
enzymes, 16, 17
epilepsy, 3, 18, 177
epinephrine. *See* adrenaline
episodic memory, 166–67
eugenics, 4
euphoria, 26, 73
evolution, 136, 148–49, 182
exercise, 20–21, 88, 89, 94, 100
exhaustion, 84
experiential responses, 163
explicit memory. *See* declarative memory
expression, gene, 21, 26, 87
extinction, response, 138–39
extroverts, 30, 32
Eyer, Joseph, 86

facial skeleton, 47, 48
false memory, 143
family history, of mood disorders, 56, 66–70
fast Fourier transformation, 43
fats (dietary), 89, 100
fear, 18, 126–40, 182
    children's expression of, 37–38, 40–41, 45, 47
    conditioning and, 127–32, 139
    memory and, 6, 124, 132, 138
    response extinction, 137–40
    subjective feelings and, 143–46
    unconscious emotions and, 133–37
    See also fight-or-flight response
feelings, 125, 143–46, 147
fever, 112
fibromyalgia, 118
fight-or-flight response
    disease and, 108, 118
    fear and, 34–35, 128–29, 133–35
    stress and, 5, 34–35, 81, 84–86, 94, 108, 128
Fischer rats, 117, 121
Flexner, Louis, 168
Flourens, Pierre, 155–56
flu, 112–13, 119
fMRI. *See* functional magnetic resonance imaging
Fox, Nathan, 44–45, 46
freeze reflex, 128, 129
Freud, Sigmund, 6, 48, 106, 135, 179, 180, 183, 194–95
Fritsch, Gustav, 158–59
frontal cortex, 137, 167
frontal lobe, 146, 156–57, 158
functional magnetic resonance imaging, 11, 15, 97, 177

functional neuroses, 106

Galen, 2, 30–31, 103, 104, 105
Gall, Franz Joseph, 154–55
Galton, Francis, 4
ganglia, 169
gastric ulcers, 83
gender, 46–47, 71–72
gene manipulation, 5, 75–76, 79
general adaptation syndrome, 82, 84
genes, 1–5
 activation and deactivation of, 20–21, 172–73, 176–77, 178
 and addiction predisposition, 22, 23, 27
 environmental interrelationship, 4–5, 22–24, 27–28
 expression of, 21, 26, 87
 function and structure, 16–18
 individual differences in, 21–24
 as mental illness factor, 9–12, 22–23, 28, 54, 56, 66–70, 72, 78–79, 176
 as personality determinants, 31, 35, 40–41, 48, 49
 protein production, 16–17, 20–21, 87, 108, 152, 171–73, 177, 178
 stress response and, 87, 99, 120–21
genetic testing, 5, 75–76, 78
genome, 16
glutamate, 19, 96–97, 171
Greek mythology, 180
growth factors (proteins), 16
guilt, 182

habituation, 84, 88
hallucination, 57, 179, 181, 189, 194, 198
health and illness, 27–28
 emotions and, 5–6, 103–21
 genetic predispositions, 22–24, 176
 mind-body connection, 2–4, 103, 104–10
 sickness behavior, 112–13
 stress effects on, 5, 82–84, 88–91, 93, 97–99, 108, 111
 See also immune system; mental illness; *specific conditions*
heart disease. *See* cardiovascular disease
helplessness, 99, 101
Hemingway, Ernest, 69
Hemingway, Margaux, 69
hemoglobin, 16
heritability. *See* genes
high-reactive behavioral profile, 35–50
hippocampus, 125, 177, 196
 memory and, 94, 137, 140–42, 145, 147, 164–65, 166–67
 stress-response role, 94–99, 100
Hippocrates, 30, 47–48
Hitler, Adolf, 4
Hitzig, Eduard, 158–59
HIV virus, 119
H.M. (patient), 164–66
homeostasis, 82, 84, 87
homicide, 83
homunculus, 174, 175
hopelessness, 99, 101
hormonal action 6, 82–88, 91–92, 94–98, 100, 104, 110–11, 113–19, 121, 128, 142 (*see also specific hormones*)
hospitalization patterns, 62–63, 71
HPA axis. *See* hypothalamic-pituitary-adrenal axis
Hubel, David, 19
humors, bodily, 2, 31, 105
Huntington's disease, 22
hypertension, 82, 83, 88, 90
hypnagogic hallucination, 194
Hypnos (god), 180
hypoalgesia, 129
hypothalamic-pituitary-adrenal axis, 85–87, 110–12, 116–19, 128, 142
hypothalamus, 125, 182
 and cortisol production, 34, 85, 142
 and fight-or-flight response, 84–85
 immune system and, 108, 110, 112, 113–15

hypothalamus *(continued)*
neuronal communication, 14
hysteria, 106

Icelus (god), 180
identity, 1, 3, 7
immune cells, 89, 91–93, 111–13
immune system, 91–93
brain and, 94, 110, 111–17
cytokines and, 88–89, 104, 111–13
emotions and, 5–6, 103–4, 110, 117–21
neurotransmitters and, 31, 104, 110
stress and, 5, 85, 87, 89, 91–93, 108, 110–11, 113–15, 119–21
implicit memory. *See* nondeclarative memory
infantile amnesia, 142
infection, 18, 116, 119
inflammatory diseases, 104, 116–19
influenza, 112–13, 119
inhibited behavioral profile, 33–34, 35, 37, 40, 44, 45, 47, 49–50
insulin, 89
intellect, 155, 156
interleukin-1, 112, 113, 115, 117
interleukin-2, 112
introverts, 30, 32, 38–39
IQ functioning, mania and, 74

Jung, Carl, 30

Kagan, Jerome, 120
keratin, 16
kinase, 170, 171
Kissinger, Henry, 20
Kraepelin, Emil, 73–74

LaBar, Kevin, 139
language, 19–20, 156–61
Lashley, Karl, 162
learning
addiction mechanism and, 26–27
brain plasticity and, 15, 19–20, 173–76, 177
emotional, 131–33, 136, 139
memory and, 6, 151–52, 167, 168, 173, 199
molecular basis of, 145
Leborgne (Broca's patient), 158
leukocytes, 110, 111–12
Lewis rats, 116–17, 121
life expectancy, 83
limbic system, 125
lithium, 76–77
locus ceruleus, 115
long-term memory, 6, 20, 137, 153, 164–73
Lowell, Robert, 55, 57, 58
low-reactive behavioral profile, 35–41, 43–45, 47–50
low road. *See* thalamo-amygdala pathway
Ludwig, Arnold, 71, 72
lymphatic system, 110
lymph nodes, 91, 110, 115

macrophages, 111
magnetic brain stimulation, 54
magnetic resonance imaging, 11, 15, 97, 177
Maimonides, Moses, 104, 105
malnutrition, 18
manic-depressive illness, 4, 10, 22, 23
in creative people, 5, 53–79
onset and episodic pattern of, 58
Marchand, Leslie, 59
mass action, law of, 162
Mayakovsky, Vladimir, 64
McLean, Paul, 125
MDI. *See* manic-depressive illness
meditation, 120
medulla oblongata, 155
melancholia. *See* depression
melancholic personality, 30, 103
melanocytes, 48

Melville, Herman, 148
memory, 123–24
  brain's storage of, 151–52, 153–56, 162–69
  conscious, 140–42, 145, 147
  declarative, 166–69
  dreams and, 6, 179, 182–83, 191, 196–97, 199
  emotional, 6, 124, 138, 140–47
  episodic, 166–67
  false, 143
  fear and, 6, 124, 132, 138
  immune cell, 91
  learning and, 6, 151–52, 167, 168, 173, 199
  long-term, 6, 20, 137, 153, 164–73
  molecular basis of, 6, 20, 153, 178
  nondeclarative, 166, 167–78
  semantic, 166–67
  short-term, 6, 153, 165–68, 170–71
  storage mechanisms, 167–69
  stress's effect on, 5, 82, 95, 96
  trauma's effect on, 142–43
  working, 137, 144, 146–47, 184
memory loss, 6, 152, 178, 182–83
Mendel, Gregor, 22
mendelian traits, 22
mental illness
  brain-function abnormalities and, 2–3, 31
  creativity and, 5, 53–79
  dreaming as model for, 183
  genetic-environmental interrelationships, 4, 9–12, 22–23, 28, 54, 66–70, 72, 78–79, 176
  See also specific types
mental retardation, 18
Merzenich, Mike, 174–75
mesomorph body type, 48
metabolic system, 87, 94, 198
Meyer, Adolph, 62
Milner, Brenda, 164–66
mind
  definition of, 3
  inseparability from brain, 4
  See also brain; memory; unconscious mind
mind-body connection, 103 (*see also* emotions)
mind-body integration, 2–4
molecular biology, 145, 152–53, 178
mood disorders, 53–79
  family-history factor, 56, 66–70
  immune system and, 93, 104, 118
  onset and episodic recurrences of, 58
  productivity patterns and, 61–63, 76–77
  rates among creative people, 70–75
  retrospective diagnosis of, 55–58, 62
  suicide potential, 5, 54–55, 58, 63–66, 69–70, 71, 72, 77
  treatment of, 75–79
  See also depression; manic-depressive illness
Morpheus (god), 180
motor cortex, 159–60
motor responses, 111, 184, 185, 191–92
MRI. *See* magnetic resonance imaging
muscles, 20–21, 185, 190
mythology, 180

"nature vs. nurture" debate, 2, 4–5 (*see also* genes; environment)
neocortex, 132
nerve cells. *See* neurons
neural crest, 48
neurobiology, 153
neurochemistry, 12–21
  addiction mechanisms, 25–27
  immune system, 5–6, 104, 110
  as personality determinant, 29–32, 34–35
  of sleep and dreams, 179, 184–85, 189, 191, 198–99
neuroendocrine immunology, 104

neurogenesis, 95–96
neurons, 19, 177
  addiction mechanism and, 25–26
  brain maps and, 129–30
  damage to, 3, 96–97, 113
  dream activity and, 179, 185, 189, 193, 195
  function and structure, 7, 12–18
  gene activation and deactivation, 20–21, 173
  nerve-cell formation, 95–96
neuroscience, 2–3, 9, 124, 178
  and study of memory storage, 154–56
  tools for studying fear, 129–33
neuroses, functional, 106
neurotransmitters
  dream activity, 179, 184–85, 189, 191, 198
  function and structure, 13–14, 16, 19, 21
  immune system, 31, 104, 110
  neuron damage and, 96–97
  as personality determinants, 31
  See also specific kinds
nightmares. *See* dreams
nondeclarative memory, 166, 167–78
non-REM sleep, 183–84, 186–88, 190, 197
noradrenaline. *See* norepinephrine
norepinephrine, 31, 84–85, 96, 184, 189, 191
NREM sleep. *See* non-REM sleep
nucleotide bases, 16
Nyx (god), 180

Obecalp, 107
obesity, 89, 90
obsessive-compulsive disorder, 3, 126
occipital lobe, 156–57
opiate peptides, 128
opiates, 107
optimism, 99
pain, 128–29
panic attacks, 126, 144
paranoia, 57
parasympathetic nervous system, 42–43, 84, 85
parietal lobe, 156–57
passion, 147–49
patent medicines, 107
Pavlov, Ivan, 120, 127, 167
Penfield, Wilder, 163
personality development, 29–51
  biological potentials and, 2, 30–32, 41–46, 48–50
  gender and cultural factors, 46–48
  genes vs. environment, 4–5, 40–41
  high- vs. low-reactivity tests, 32–40
  sensory feedback and, 50–51
PET. *See* positron-emission tomography
Phantasus (god), 180
pharmaceuticals. *See* drugs
Phelps, Liz, 139
phlegm (humor), 2, 31, 105
phobias, 138–39, 144, 145
phrenology, 155
pituitary gland, 85
placebo effect, 107–8
plasticity, brain, 3, 7, 14–20, 94, 177
Plato, 148
Poe, Edgar Allan, 62–63, 65–66
poetry writing, manic-state, 73–74
pons, 184–85, 189–91, 193
positron-emission tomography, 11, 15, 73, 108, 177
Post, Felix, 71, 72
post-traumatic stress disorder, 98, 99, 106, 126
prefrontal cortex, 137–38, 139
pregnancy, 17, 18, 42
production-line study, 100
prostaglandins, 113
proteins
  addiction mechanism and, 25–26
  consolidation phase, 168–69

cytokines, 88–89, 104, 111–12
gene production of, 16–17, 20–21, 87, 108, 152, 171–73, 177, 178
memory transition and, 167–71, 178
psychosomatic medicine, 106
psychotherapy, 106, 145, 176

Ramon y Cajal, Santiago, 7
rationality, dreams and, 181, 182, 189
reason, 147–49
receptors, 14, 16, 21, 113
Rechtschaffen, Allan, 197–98
reflexes, 127–29
relaxation therapy, 120
REM sleep, 183–92, 195, 196–99
Renoir, Pierre-Auguste, 119
repressors, gene, 172–73, 178
resistance, adaptation phase, 84, 85
retina, 17, 19
rheumatoid arthritis. *See* arthritis
Richards, Ruth, 72
risk taking, 49
Roethke, Theodore, 57, 62
Russia, 83

SAD (seasonal affective disorder), 118
sadness, 182
sanguine personality, 30, 103
Sapolsky, Robert, 97
schizophrenia, 10
brain abnormalities and, 3, 18, 98
creativity and, 70–71, 74
dreaming as model for, 183
genetic component, 22, 23, 176
Schumann, Robert, 62, 63, 69
Scoville, William, 164
seasonal affective disorder, 118
"second hits" (environmental factors), 4, 10, 24
selective breeding, 4
self, 1, 3
Selye, Hans, 82, 83, 84, 108
semantic memory, 166–67
senile dementia, 59
sensitization, 170
sensory cortex, 137
sensory systems, 18–19, 20
fear learning and, 131–32, 136–37
feedback effects on personality, 50–51
immune cells and, 111–13
serotonin, 14, 31, 184, 189, 191
Sexton, Anne, 69
shame, 182
shell shock, 106
Sherrington, Charles, 163
short-term memory, 6, 153, 165–68, 170–71
shyness, 30, 32–41
sickle-cell anemia, 22
sickness behavior, 112–13
Slater, Eliot, 62
sleep, 7, 179–99
manic-phase reduction of, 61
stress regulation and, 100
See also dreams
sleepwalking, 190
smoking, 24, 100
socialization, 47, 49
social support, 100, 121
Socrates, 148
sound perception, 131–32
spikes. *See* action potentials
spleen, 91, 110, 115
Squire, Larry, 168
startle reflex, 128
stereotypes, cultural, 47
Sterling, Peter, 86
stress, 81–101
allostasis and allostatic load, 86–89, 93, 99, 100
brain function and, 5, 82, 84–85, 94–99, 108–10, 115
cardiovascular system and, 87, 88, 89–91
coping with, 84, 94, 99–101
fear memories reactivated by, 138

stress *(continued)*
fight-or-flight response to, 5, 34–35, 81, 84–86, 94, 108, 128
good vs. bad, 83–84
immune system and, 5, 85, 87, 89, 91–93, 108, 110–11, 113–15, 119–21
See also fear
strokes, 3, 43–44, 91, 157
structural magnetic resonance imaging, 11
substance abuse, 10, 18, 77
genetic–environmental interactions, 4, 22, 23–24
mechanism of addiction, 24–27
REM-sleep suppression, 198
suicide
among creative people, 5, 54–55, 58, 64–66, 69–72, 77
stress factor, 83
Swift, Jonathan, 199
symbolism, dream, 7, 195–96
sympathetic nervous system
behavioral research, 34–35, 41–43, 46, 48, 50
fight-or-flight response and, 34–35, 84, 85
immune system, 115
synapses, 13–14, 18
addiction mechanism and, 25–26
learning and, 15, 20, 173, 177
remodeling process, 20

temperament. *See* personality
temporal lobe, 94, 95, 156–57, 160, 163–65, 166
Tennyson, Alfred, 66–67, 68
tension, 50–51
thalamo-amygdala pathway, 132–37, 144
thalamus, 17, 19, 125, 131–35, 136
Thanatos (god), 180
Thomas, Edward, 56
thymus, 91, 110, 115
trafficking of immune cells, 91–93
trauma, 99, 142–43
Twain, Mark, 1
twin studies, 22–23

uinpolar depression, 58
ulcers, gastric, 83
unconscious mind
dreams and, 179, 194–95
emotions and, 6, 135–36, 137, 140–43, 147
uninhibited behavioral profile, 33–34, 35, 37, 40, 44, 45, 47, 49

vagus nerve, 112
van Gogh, Cornelius, 70
van Gogh, Theo, 70
van Gogh, Vincent, 62–63, 64, 70
van Gogh, Wilhelmina, 70
visual cortex, 17, 19, 189
visual sensory system, 18–19
volition, 2, 155, 156

war neurosis, 106
Weisel, Torsten, 19
Wernicke, Carl, 160–61
Wernicke's area, 159, 160, 161
white blood cells (leukocytes), 110, 111–12
Whitehall Studies, 90
will. *See* volition
withdrawal behaviors, 45, 198
Wittgenstein, Ludwig, 51
Woolf, Virginia, 69
working memory, 137, 144, 146–47, 184

yellow bile (humor), 31, 105